Community-based Adaptation to Climate Change

Praise for this book

'A long overdue critical analysis of community-based adaptation, linking conceptual challenges with vibrant examples from practice. This book will be fascinating reading for students, researchers, policy makers, and development workers interested in tackling the impacts of climate change from the ground up.'

Thomas Tanner, Overseas Development Institute, UK

'A fascinating and clear explanation of the objectives and challenges of community-based adaptation. I especially liked the case studies and valuable lessons for policy makers. This book will be an invaluable guide to students and practitioners who want to implement climate change policy in developing countries.'

Tim Forsyth, London School of Economics and Political Science

'Action to support those most vulnerable to the impacts of climate change is more urgent than ever. This timely publication with its wealth of practical experience shows how it can be done, what sort of challenges are likely to emerge on the way, and crucially, how it can be scaled up. In particular, the emphasis on the central importance of environmental sustainability and equity to current and future adaptation and resilience make it essential reading for practitioners and policy makers alike.'

Richard Ewbank, Climate Advisor, Christian Aid

'The burden of climate change will be felt most keenly and with greatest impact in the real lives and experiences of people and communities around the world. There is no more important question than how communities can adapt. It is a matter of justice, of power and with living with nature. This book illuminates that landscape and points the way to how to use adaptation in a proactive and progressive manner. This is important research.'

Neil Adger, University of Exeter, UK

'The actions of individuals, households and communities will be critical in responding to the challenges of climate change. The chapters in this book provide a thorough analysis of how community-based adaptation has already been contributing to this process, and point to some useful directions for how it can do so even more effectively in the future. The contributors and editors bring together the key strands of research, policy development, and project implementation, and provide a valuable resource for people seeking to understand and support CBA around the world.'

David Dodman, Senior Researcher, Human Settlements and Climate Change, International Institute for Environment and Development

Community-based Adaptation to Climate Change
Emerging lessons

Edited by
Jonathan Ensor, Rachel Berger and Saleemul Huq

PRACTICAL ACTION
Publishing

Practical Action Publishing Ltd
The Schumacher Centre
Bourton on Dunsmore, Rugby,
Warwickshire CV23 9QZ, UK
www.practicalactionpublishing.org

ISBN 978-1-85339-790-5 Hardback
ISBN 978-1-85339-791-2 Paperback
ISBN 978-1-78044-790-2 Library Ebook
ISBN 978-1-78044-791-9 Ebook

A catalogue record for this book is available from the British Library.

Ensor, J., Berger, R. and Huq, S. (2014) *Community-based Adaptation to Climate Change: Emerging Lessons*, Rugby, UK: Practical Action Publishing <http://dx.doi.org/10.3362/9781780447902>.

Since 1974, Practical Action Publishing has published and disseminated books and information in support of international development work throughout the world. Practical Action Publishing is a trading name of Practical Action Publishing Ltd (Company Reg. No. 1159018), the wholly owned publishing company of Practical Action. Practical Action Publishing trades only in support of its parent charity objectives and any profits are covenanted back to Practical Action (Charity Reg. No. 247257, Group VAT Registration No. 880 9924 76).

Cover photo: A member of the Association of Innovative Farmers of Zandoma, Burkina Faso, digs *zai* planting holes, one of a number of agricultural practices that have enhanced resilience on their farms.
Photo credit: M. Tall, CGIAR Research Program on Climate Change, Agriculture and Food Security, West Africa
Cover design by Mercer Design
Indexed by Liz Fawcett, Harrogate, UK
Typeset by Allzone Digital

Printed in the United Kingdom by Hobbs the Printers Ltd, Totton, Hampshire

FSC www.fsc.org MIX Paper from responsible sources FSC® C020438

Contents

http://dx.doi.org/10.3362/9781780447902.000

Photographs, figures, tables and boxes

Photographs

Figures

Tables

Boxes

About the editors

Jonathan Ensor is a lecturer at the Centre for Applied Human Rights, University of York. He is author of *Uncertain Futures: Adapting Development to a Changing Climate* and co-author with Rachel Berger of *Understanding Climate Change Adaptation.*

Rachel Berger was formerly the climate change policy adviser at Practical Action and now works independently on climate change adaptation issues.

Saleemul Huq is Director of the International Centre for Climate Change and Development at the Independent University, Bangladesh, and a Senior Fellow, International Institute for Environment and Development. He is a lead author of the Intergovernmental Panel on Climate Change.

Acronyms and abbreviations

ALRMP	Arid Lands Resource Management Project
ARCAB	Action Research for Community Adaptation in Bangladesh
ASAL	arid or semi-arid land
CATIE	Centro Agronómico Tropical de Investigación y Enseñanza
CBA	community-based adaptation
CBD	Convention on Biological Diversity
CBNRM	community-based natural resource management
CBO	community-based organization
CCF	community capitals framework
CDM	clean development mechanism of the United Nations
CFR	Cape Floral Region
CGS	Council for Geoscience South Africa
COP17	17th Conference of the Parties (UNFCCC)
CP	co-operating partner
CRD	Center for Rural Development in Central Vietnam
CRiSTAL	Community-based Risk Screening Tool – Adaptation and Livelihoods
CSIR	Council for Scientific and Industrial Research
DDC	district development committee
DMC	disaster management committee
DPSC	district project steering committee
DRR	disaster risk reduction
DSG	district steering group
EbA	ecosystem-based approaches to adaptation
EMG	Environmental Monitoring Group
FDP	final distribution point
FFA	Food for Assets
GoK	Government of Kenya
HKH	Hindu Kush Himalaya
ICIMOD	International Centre for Integrated Mountain Development
ICRAF	World Agroforestry Centre
IDS	Institute of Development Studies
IFAD	International Fund for Agricultural Development
IIED	International Institute for Environment and Development
IPCC	Intergovernmental Panel on Climate Change
KFSSG	Kenya Food Security Steering Group
KRCS	Kenya Red Cross Society
LAC	local adaptive capacity
LDC	least developed country

MSME	micro, small and medium-scale enterprise
NALEP	National Agriculture and Livestock Extension Programme
NAPA	National Adaptation Programme of Action
NGO	non-governmental organization
NWP	Nairobi Work Programme
PAR	participatory action research
PRA	participatory rural appraisal
RC	relief committee
RNFE	rural non-farm economy
RRA	rapid rural appraisal
SEI	Stockholm Environment Institute
TVE	township and village enterprise
UCT	University of Cape Town
UNEP	United Nations Environment Programme
UNFCCC	United Nations Framework Convention on Climate Change
UNWFP	United Nations World Food Programme
USB	University of Stellenbosch
VCA	value chain analysis
VCG	value chain group
VDC	village development committee

CHAPTER 1

Introduction: Progress in adaptation

Rachel Berger and Jonathan Ensor

The global context for community-based adaptation (CBA) is of an increasing certainty of a global temperature rise greater than 2 degrees Celsius by the end of this century, and an absence of adequate action on mitigation. In some parts of the world, the severity of environmental impacts may exceed the conditions for effective adaptation. The chapter reviews the progress in understanding of CBA and the emerging literature from research as well as from field programmes. It defines CBA, and also adaptive capacity, a key element in adaptation, and looks at conceptual frameworks for understanding CBA. The importance of social networks, diversification and innovation and the power relations inherent in access to resources are briefly examined, concepts that are developed more fully in the thematic chapters. Brief introductions to the themes of the case study chapters follow, and the chapter concludes with some outstanding challenges for researchers, practitioners and policy makers.

Keywords: community-based adaptation, climate change, definitions, networks and conferences, adaptive capacity, transformation, scaling up

The global context

This book is being published at a time of paradox. The science is becoming explicit about the certainty and speed with which the earth's climate is changing, and the rapidly diminishing time available for taking action to prevent dangerous climate change. If action to reduce greenhouse gas emissions is not taken rapidly, global mean surface temperatures are projected to continue to increase, and there is a high degree of probability of the rise exceeding 2 degrees Celsius above pre-industrial levels by the end of this century. We therefore have very few years in which to take the drastic action necessary to stabilize the climate, and yet the probability of timely international action commensurate with the need is small; in the most powerful countries, action on climate change is low on the political agenda compared with economic growth and securing energy supplies.

In 10 years we have moved from a time when it was possible for some to believe that mitigation (reduction of emissions of greenhouse gases) would be sufficient action, and adaptation to climate change would be needed only in certain places, to the realization that in many parts of the world adaptation will soon no longer be possible, because of the severity of the environmental changes consequent upon climate change – changes such as sea-level rise,

http://dx.doi.org/10.3362/9781780447902.001

longer droughts and increased frequency of flooding. In these 10 years, the realization that the impact of climate change falls most heavily on populations that are already vulnerable, and who have contributed least to the problem, has led to the growth of activity – academic research, action research by non-governmental organizations (NGOs) and internationally funded programmes of on-the-ground work – to support communities in understanding the challenges that climate change is presenting, and in adapting to them.

Defining community-based adaptation

Community-based adaptation (CBA) to climate change is a community-led process, based on communities' priorities, needs, knowledge and capacities, which seeks to empower people to prepare for and cope with the impacts of climate change (Reid et al., 2009). CBA comprises several components and is rooted in participatory development programmes to strengthen livelihoods and reduce vulnerability, as well as disaster risk reduction thinking to build resilience to climate-related disasters. As Reid and colleagues explain, 'CBA needs to start with communities' expressed needs and perceptions, and to have poverty reduction and livelihood benefits' (2009). Several components are specific to the challenge of dealing with future climate change: the need for scientific information on seasonal forecasts, long-term climate predictions, and local trends in temperature and rainfall, combined with local knowledge and previously successful coping strategies (Ensor and Berger 2009). Climate change modelling embodies considerable uncertainty both about the predictability of longer-term changes in the face of possible tipping points such as the melting of permafrost and Arctic sea ice, and about the changes at sub-regional levels. Dealing with uncertainty is thus a crucial element of CBA and involves building capacity to make decisions that will minimize the risk to livelihoods and assets of an extreme event, and building in an ability to cope with constant change. Dealing with uncertainty and change is likely to involve experimentation and innovation, developing new ways of producing food and earning a living. Another key component of CBA is access to information and knowledge – whether scientific knowledge about new crop varieties, or technologies that will use scarce resources more effectively. Knowledge is power, and poor people are marginalized because they lack power over key resources – including land, water and information. CBA must therefore assist people to organize effectively to participate in local decision-making processes and to mobilize to challenge situations that affect their ability to adapt to climate change.

'Community-based adaptation' is still a relatively new concept, not widely known outside the development community. As a term it has a cosiness, a universal acceptability – who could possibly disagree with a set of interventions aimed at helping poor people cope with current and likely future challenges posed by the changing climate? Buried in this term, though, are two words, the meanings of which need unpacking and examining for the assumptions and concepts embedded within them if there is to be real and long-lasting support for vulnerable people in addressing the uncertain futures that climate change will bring. 'Community' is a term used to describe a group of people with common interests, and in development it is generally used to describe people living in the same village or area. It assumes that people in that particular locality will face shared challenges and therefore will have an interest in working jointly to address those challenges. Place is an important factor in determining vulnerabilities due to climate change, but other factors may have as great, or greater, impact. Even in the smallest villages there is diversity of resources and skills. Within households there will be customary roles and values that will lead to some members being disadvantaged in accessing information, skills and resources that would help them address existing and new challenges to their livelihoods. Access to the resources needed for adaptation is not just about their physical location but about power relations both within a community and beyond. Communities are also not merely place-related: people are members of several communities, depending on age, gender, employment, cultural groupings, and so on. Support and skills for adaptation may be drawn on from these other communities. Unpacking the meaning of 'community' and the extent to which adaptation in one location will involve activity across scales of governance is a theme considered by Yates in Chapter 2 and returned to in the conclusion.

In recent years, several conceptual frameworks have been developed for CBA, and most of them include the elements noted above. The framework put forward in Ensor and Berger (2009) and further developed in Ensor (2011) underpins this book, describing adaptation in terms of vulnerability reduction, absorbing capacity and adaptive capacity. These terms refer to the different ways in which adaptation efforts may address the challenges of climate change: by aiming to reduce vulnerability to particular impacts, such as flooding; by improving the capacity to cope with a broad array of threats, such as through livelihood diversification; and by increasing the ability of households and communities to make changes in response to experienced or anticipated climate change. The Intergovernmental Panel on Climate Change (IPCC) defines adaptive capacity as 'the property of a system to adjust its characteristics or behaviour, in order to expand its coping range under existing climate variability, or future climate conditions' (Brooks et al., 2005). While adaptation can be undertaken as a response to a climate-related problem, enhancing the adaptive capacity of local communities to climate change implies an approach to adaptation that is forward looking.

Adaptive capacity takes adaptation beyond reduction of vulnerability to hazards and disaster preparedness, involving an ongoing change process where communities can make decisions about their lives and livelihoods in a changing climate (Ensor, 2011). Adaptive capacity must also focus on addressing the political, cultural and socioeconomic factors that may promote or inhibit individuals and groups from adapting (Smit and Pilifosova, 2001). Institutions and governance have therefore increasingly been recognized as central to adaptive capacity (Eakin and Lemos, 2010; Engle, 2011; Gupta et al., 2010), directing attention both to formal decision-making processes and to the role of social norms and practices in guiding choices and actions. In Ensor (2011), adaptive capacity is considered in terms of the processes that must be in place if communities are to be able to make changes to their lives and livelihoods in response to emerging environmental change. Acquisition of skills, resources and information is dependent in turn on access to and control of the material and knowledge assets that will determine the range of options that communities have at their disposal. As a result, three broad and interlinked dimensions are seen as central to development efforts to secure adaptive capacity: addressing power relationships and decision-making institutions; the control, distribution and generation of knowledge; and opportunities for experimentation and innovation. While adaptive capacity was given limited attention in the case studies reported by Ensor and Berger (2009), each of these areas is addressed in different ways by the contributors to this volume. As discussed in the concluding chapter, attention is directed to relationships and networks; the distribution of power, knowledge and influence; and the implications of culture and gender. This view also invites CBA practitioners to engage with different geographical scales, meaning that for effective adaptation local communities may need to link or engage with actors and institutions located in different and sometimes distant locations.

Community-based adaptation: the current state of play

CBA has grown from pilot testing of new approaches to an emerging field of development studies embracing conceptual frameworks, governance, power structures, change and uncertainty, all in the context of underlying poverty, vulnerability and inequity of resource distribution and access. Academic interest and NGO programme activity in the field of adaptation are increasing rapidly, with the establishment of postgraduate programmes in climate change and development at several universities, including the International Centre for Climate Change and Development at the Independent University of Bangladesh. That department is partnered with 11 international NGOs in a programme called Action Research for Community Adaptation in Bangladesh, involving 20 field sites in different ecological zones. Knowledge and experience gained in Bangladesh will be transferred to other countries where the partners operate.

There is an increasing body of followers, practitioners and researchers with an academic and professional interest in CBA, and an increasing awareness of the field among developing country governments. In 2010 the first conference to focus on adaptation science was held in Australia. It brought together around 1,000 climate change adaptation experts and practitioners from NGOs, research institutions, universities and professional organizations from over 50 countries. The themes covered included understanding and communicating adaptation, adaptation in different sectors, grassroots case studies, frameworks for adaptation, and human welfare aspects of adaptation (community, social, equity, health). A key challenge highlighted was the multidisciplinary nature, encompassing livelihoods, infrastructure, disaster risk reduction, economics, food security, ecosystems and sustainable development. A second conference in the series was held in Arizona in 2012, focusing on adaptation to climate variability and change, and looking at equity and risk, learning, capacity building, methodologies, adaptation finance and investment, and ecosystem-based adaptation approaches. It explored practical adaptation policies and approaches, and strategies for decision making from the international to the local scale. These conferences are only two high-profile examples from the many gatherings that have touched on the challenges presented by climate change. Their scale, however, is illustrative of the growing level of academic, government and NGO interest in the field as well as the breadth of issues that are encompassed by the term 'adaptation'.

Despite the growing awareness of the need for adaptation both in developing and developed countries, it is an understatement to say that adaptation is seriously underfunded. While adaptation is integrally linked to development, the World Bank estimates that the cost of adaptation in developing countries will be about $75 billion to $100 billion by 2020, above and beyond the baseline cost of development (World Bank, 2010). Recent studies suggest that developing countries will need funds in the range of $100 billion to $450 billion a year for adaptation actions. Currently, funding totals are a fraction of this figure – $2.6 billion pledged as of November 2012, but $1.9 billion delivered so far (Schalatek et al., 2012). The funds available flow through a number of channels, including the European Union's Global Climate Change Alliance, and the United Nations Framework Convention on Climate Change's (UNFCCC's) Least Developed Countries Fund and the Special Climate Change Fund. Recently, the UNFCCC's Adaptation Fund and the Pilot Program for Climate Resilience of the World Bank's Climate Investment Funds have added significantly to the funds approved.

While international funding for climate change adaptation is still woefully inadequate, an increasing number of developing countries are beginning to plan for adaptation to climate change, in recognition of the fact that many of the resources needed will have to come from a shift in allocation and prioritization of governments' own budgets, and from private investment. Among vulnerable countries, Bangladesh and Kenya already have a climate change strategy and Nepal has one in preparation. The UNFCCC process set up an Adaptation Committee at COP17 in Durban, which has met and has begun its programme of work. This includes offering guidance on preparing national adaptation plans. The UNFCCC Adaptation Fund has been meeting since 2008 to disburse funds for projects and programmes that are submitted by developing country governments and that meet the Fund's criteria.

Because of the poor outlook for an adequate global agreement to prevent dangerous climate change, there is a critical need for considerable effort to be put into research that will help those most vulnerable to the adverse impacts of climate change. In working with poor and vulnerable people, one objective must always be to seek to address the pressing needs of people struggling to cope now. While the sustainability of CBA activities, in terms of relevance and durability over the longer term, may well be challenged by more rapid climate change or more extreme events than could be foreseen or addressed, there is no doubt about the urgency and importance of CBA in the coming years. Currently, adaptation funds are allocated to a range of adaptation activities, including large-scale infrastructure such as dams and sea walls, national planning for adaptation and disaster preparedness, and to many countries that are not among the poorest.[1] Within the countries receiving funds, the most vulnerable communities are unlikely to be seeing a direct benefit, since CBA is rarely a national priority. This may be beginning to change. The Adaptation Fund has a set of criteria for project appraisal, which include addressing specific climate-related vulnerabilities and involving a wide variety of stakeholders, including affected communities, in both the design of the programme or project and its implementation. Nepal is leading the way by looking to implement adaptation through local adaptation programmes of action.

In support of CBA, knowledge networks on adaptation are developing apace. In the past few years, weADAPT has developed into a significant knowledge platform on adaptation. The Regional Climate Change Adaptation Knowledge Platform for Asia (Adaptation Knowledge Platform), a development led by the United Nations Environment Programme (UNEP), the Asian Institute of Technology and the Stockholm Environment Institute, has been developed to respond to demand for effective mechanisms for sharing information on climate change adaptation and for developing adaptive capacities in Asian countries. The UNFCCC Nairobi Work Programme (NWP) is a capacity-building and knowledge-sharing programme on vulnerability, impacts and adaptation. While CBA currently barely features on the website, this is set to change as the NWP has committed to making its large body of information more readily available to non-experts and to the local level so that community groups and local NGOs can access information in a usable form, such as podcasts, short videos or technical briefs on adaptation technologies and approaches suitable for community-based action.

Supporting research and practice in CBA are the international conferences on CBA organized by the International Institute for Environment and Development (IIED) along with academic, NGO and government partners in Bangladesh, Tanzania and Vietnam since 2004. These conferences have provided a valuable forum for academics and practitioners to present their work to peers for discussion, and an encouragement to document good practice and innovative approaches. A platform for sharing information, the Global Initiative for CBA, was established in 2009 at the third conference on CBA in Dhaka, bringing together a large number of practitioners in a loose network. Several web portals seek to provide easy access to case studies of good practice on adaptation, including weADAPT.

The aims and structure of the book

With growing action on adaptation, there is a need for literature that provides reflections on key issues such as the nature of adaptation, the challenges and pitfalls of participatory approaches in the context of CBA, the need for transformative approaches, looking at options other than purely adaptation finance and assessing the economic benefits of adaptation interventions. This book contributes to filling this need. It follows on from work by Ensor and Berger (2009) and Ensor (2011), and a recent book on scaling up from projects on CBA (Schipper et al., 2014). There are now a significant number of case studies, many based on doctoral research, looking at field programmes that address adaptation to climate change either as a main objective or as a co-benefit with other development issues. This book goes beyond being a collection of case studies of small-scale projects to raise some issues that must be resolved if CBA is truly to provide viable and sustainable solutions for vulnerable communities.

The first part of the book contains three thematic chapters addressing some important issues not widely touched on in CBA literature. First, Yates critically examines the meaning of 'community' and questions the assumption that local is always the appropriate scale even for CBA. As Dodman and Mitlin argue (2013), CBA practitioners need to engage with issues of power and governance, and not focus exclusively on the reduction of poverty and vulnerability. CBA needs to include tools and methods that enable a more explicit transfer of power to local communities and engagement with the state at multiple levels, so that decisions affecting a community are not made largely by those outside that community. Rent seeking by powerful groups is a feature of life that will increase as climate change offers new opportunities for those with power and resources. Yates argues similarly, and discusses how too strong an emphasis on the local can underplay power relations and inequalities, and therefore the need for extra-local networks and support.

In the second thematic chapter (Chapter 3), Reid explains the difference between CBA and ecosystem-based approaches to adaptation (EbA) and the importance for practitioners of fully taking account of ecosystems in developing adaptation options. At the same time as people will need to adapt their livelihoods, the environment – comprising soil, water resources, plants, animals, microfauna and microflora – will also be adapting autonomously in ways that need to be understood so that human activities that depend on natural resources take full account of and work in synergy with these changes in the local ecology. EbA may be especially important to consider for vulnerable rural communities, since their livelihoods are highly dependent on natural resources.

To date, adaptation has generally been viewed as involving relatively minor changes to livelihood patterns and local natural resource management. Most field programmes on adaptation have been in rural areas, and while livelihood options less dependent on natural resources have often formed part of an adaptive future, this has not been seen as the main adaptation priority. In the third thematic chapter (Chapter 4), Cannon explores whether this should change, recognizing that in some places climate change will make natural resource-based livelihoods unviable. He presents case studies from China and India showing how widespread transformation of rural livelihoods has been achieved in these two countries, but in very particular circumstances and with generous financial resources. He puts forward an argument for adaptation funding being channelled to support large-scale diversification into non-farm-based livelihoods.

The six case studies that follow the thematic chapters each highlight a particular aspect of CBA. Imbach and Prado (Chapter 5) test a framework for examining how people in two villages in Mexico perceive the impact of climate change on their lives, and their response to different adaptation options. By observing whether communities respond actively or passively, the framework indicates whether programmes are building adaptive capacity. A study of the planning implications for adaptation in the Himalayan regions of four countries (Pakistan, India, China and Nepal) is presented by Pradhan, Khadgi and Kaur in Chapter 6 with reference to policies on water, agroforestry and migration. They discuss the interface between policy designed to facilitate climate change adaptation and the institutions that implement or operate within that policy framework. They highlight cases where policy has been consultative and supportive of local institutions as a channel for implementation, and cases where this has not occurred. The importance of recognizing informal as well as formal institutions and of incorporating local knowledge into policy formulation is stressed.

Chambwera and Mohammed look in Chapter 7 at the costs and benefits of a programme in Sudan that contains some key adaptation elements, offering a useful platform for considering the economics of adaptation. Since funding for adaptation will be in competition with funding for other development activities, measuring the costs and benefits and cost-effectiveness of CBA programmes will be a tool that governments will increasingly wish to use. In this case, the reach of the programme was beyond the original target number of beneficiaries and so could be seen as having potential for scaling up. From this basis, the authors develop a tool for scaling up the costs of the pilot programme from the local to the state level. Based on the cost sharing between stakeholders in the local programme, they suggest how funding for the state-level programme might be apportioned between different stakeholders, including the state government.

Koelle and Waagsaether (Chapter 8) discuss a transformative approach to building capacity among a group of farmers in South Africa producing organic rooibos tea. The importance of long-term engagement from NGOs in building capacity for learning and transformation is highlighted, presenting significant lessons for adaptive capacity building. Farmer-led research and engagement in international fair trade emerged as outcomes from the communities' own priority setting and were enabled by the communities' increased capacity.

In Chapter 9, Orindi, Mbuvi and Mutiso set out how a government food security programme funded by official development assistance has been developed into a programme that contributes to building adaptive capacity through supporting organizational capacity and increased access to information and knowledge on how to strengthen livelihoods. Given the lack of funding available specifically for adaptation, the creative and constructive use of funding for related activities, such as disaster preparedness and recovery, is increasingly necessary and it is in this regard that this case study offers useful lessons.

CBA champions stress the importance of beginning with local knowledge and experience as a basis for developing adaptation strategies. Sen and Phuong (Chapter 10) look at the role of indigenous knowledge in CBA in villages in coastal Vietnam, and how it has been integrated with scientific knowledge, and the important role this has in helping address uncertainty. There is always likely to be a high degree of uncertainty at the subnational level around the specific changes in climate to be expected in an area. Local knowledge, particularly when linked with scientific knowledge, can help fill some information gaps, enabling farmers to take action that will facilitate adaptation without unintended consequences (termed 'no-regrets' action).

Finally, in the concluding chapter (Chapter 11), Ensor draws out common themes that emerge from the case studies, suggesting important lessons for practitioners, scholars and policy makers concerned with supporting communities in their continuing efforts to address the challenges of climate change and development. The value of participation, the nature of adaptive capacity and the role of institutions in particular are identified from within the contributions. In the second half of the chapter, challenges to CBA are identified – in effective and equitable implementation, and in its scope and ambition. Reflecting on the opening thematic chapters and the case studies, Ensor discusses three areas of neglect – politics of scale and technology, ecosystems, and transformation – that CBA must come to grips with if it is to deliver on its potential and promises.

Reflections on community-based adaptation

As practitioners, we recognize that the scalability of CBA is crucial: while, to be successful, adaptation must be context specific, rooted in the local environment and culture, the huge numbers of people who will need urgent support to adapt means that adaptation programmes must move beyond NGO-funded work. Programmes for CBA must make cross-scalar linkages in order to show how CBA can be incorporated into policy frameworks for national adaptation planning. In addition, as Chambwera and Mohammed stress, enabling conditions are important – including a supportive policy framework and equitable land tenure – as are infrastructure and access to information on weather, climate predictions, technology options and markets. While scaling up is not the focus of this book, all the case studies offer lessons on these issues applicable beyond their specific local contexts.

The term 'adaptation' in general parlance is used to imply a gradual process of change, and CBA was originally framed in the late 1990s and early 2000s when containing global temperature rise to within 2 degrees seemed not only feasible but likely. In the absence of mitigation action, climate scientists are now generally discussing a rise of between 4 and 6 degrees by 2100, and even a 3 degree rise by 2050 – within the lifetimes of the majority of people alive today. The speed of environmental change in these circumstances is unmatched in any previous period of severe climatic change, whether of ice age or of global warming. Survival for many communities will depend not on an incremental approach to adaptation but on transformation. In working in the field of climate change adaptation, practitioners and researchers must address the following question: how far can the current framing of adaptation take us along the current trajectory of climate change? There is a growing literature on transformation (see O'Brien, 2012; Dodman and Mitlin, 2013; and Chapter 11) seeking to address the issue that the adaptive challenge of climate change must question the values, structures and economic systems that are driving anthropogenic climate change as well as social vulnerability and environmental degradation. Globally, the failure to tackle climate change is due to unwillingness by governments and multinational corporations, and by people as consumers, to move away from a business-as-usual development path predicated on economic growth and further depletion of the earth's resources. While this book cannot and does not seek to set out alternative models, the thematic chapters and case studies in different ways offer, at the local level, approaches and tools for adaptation that may become islands of positive change. The book is a contribution to help practitioners address the short-term challenges for adaptation. It is to be hoped that by doing so, and with the increased adaptive capacity that results, communities will be enabled to find transformative solutions to the greater medium-term challenges that climate change will throw their way.

In this light, addressing long-standing inequalities and issues such as tenure security are important for adaptation but will not always succeed because of the difficulties in resolving these issues within current systems and institutional frameworks (see Chapter 4). Challenges also exist in how to address climate change threshold effects that might render areas and livelihoods unviable, and in how to take account of migration in CBA, if that is a chosen adaptation option. These issues are cross-scalar and often cut across national boundaries. As editors we recognize the potential of community-based approaches to limit the perspective to the local scale. However, as the case studies suggest, it remains viable and important to include community-based approaches as a cornerstone in cross-scale climate responses, building *from* rather than *for* communities in addressing the wider challenges of adaptation (as suggested by Yates in Chapter 2).

Finally, adaptation at whatever scale depends on public goods – genetic material for drought-resilient crops, access to water, weather forecasting and climate predictions, and even funding – whose availability is often determined by national or international institutions such as watershed management policies. The role of the state and of international institutions is therefore critical for CBA to be successful. We hope that this book will contribute to knowledge and thinking among researchers and practitioners not only in NGOs but also in governments, where responsibility ultimately lies for ensuring that communities not only survive but flourish as climate change takes hold.

Note

1 A lack of transparency and reporting on recipients makes it very difficult to accurately assess the actual amounts flowing to vulnerable countries.

References

Brooks, N., Adger, W.N. and Kelly, P.M. (2005) 'The determinants of vulnerability and adaptive capacity at the national level and the implications for adaptation', *Global Environmental Change* 15(2): 151–63.

Dodman, D. and Mitlin, D. (2013) 'Challenges for community-based adaptation: discovering the potential for transformation', *Journal of International Development* 25: 640–59 <http://dx.doi.org/10.1002/jid.1772>.

Eakin, H. and Lemos, M.C. (2010) 'Institutions and change: the challenge of building adaptive capacity in Latin America', *Global Environmental Change* 20(1): 1–3.

Engle, N.L. (2011) 'Adaptive capacity and its assessment', *Global Environmental Change* 21(2): 647–56.

Ensor, J. (2011) *Uncertain Futures: Adapting Development to a Changing Climate*, Rugby, UK: Practical Action Publishing.

Ensor, J. and Berger, R. (2009) *Understanding Climate Change Adaptation: Lessons from Community-based Approaches*, Rugby, UK: Practical Action Publishing.

Fransen, T. and Nakhooda, S. (2012) 'Shedding light on fast-start finance' <http://insights.wri.org/open-climate-network/2012/05/shedding-light-fast-start-finance> [accessed 31 March 2013].

Gupta, J. et al. (2010) 'The adaptive capacity wheel: a method to assess the inherent characteristics of institutions to enable the adaptive capacity of society', *Environmental Science and Policy* 13(6): 459–71.

IIED (2012) 'Rich nations fail to meet 8 climate-finance pledges analysis shows' <www.iied.org/rich-nations-fail-meet-8-climate-finance-pledges-analysis-shows> [accessed 12 August 2013].

IPCC (2013) 'Summary for policymakers', in *Climate Change 2013: The Physical Science Basis. The Contribution of Working Group 1 to the Fifth Assessment Report of the Intergovernmental Panel on Climate Change*, draft report, Geneva: Intergovernmental Panel on Climate Change (IPCC).

O'Brien, K. (2012) 'Global environmental change II: from adaptation to deliberate transformation', *Progress in Human Geography* 36(5): 667–76.

Reid, H., Alam, M., Berger, R., Cannon, T. and Milligan, A. (eds) (2009) *Community-based Adaptation to Climate Change*, Participatory Learning and Action No. 60, London: IIED.

Schalatek, L., Nakhooda, S., Barnard, S. and Caravani, A. (2012) 'Climate finance for the Middle East and North Africa: confronting the challenges of climate change', *Climate Finance Policy Brief, November 2012*, London: Overseas Development Institute.

Schipper, E.L.F., Ayers, J., Reid, H., Huq, S. and Rahman, A. (2014) *Community Based Adaptation to Climate Change: Scaling it up*, London: Routledge.

Smit, B. and Pilifosova, O. (2001) 'Adaptation to climate change in the context of sustainable development and equity', in McCarthy, J.J., Canzianni, O.F., Leary, N.A., Dokken, D.J. and White, K.S. (eds), *Climate Change 2001: Impacts, adaptation, and vulnerability. Contribution of Working Group II to the Third Assessment Report of the Intergovernmental Panel on Climate Change*, Cambridge: Cambridge University Press.

weADAPT (no date) [website] <www.weadapt.org>.

World Bank (2010) *Economics of Adaptation to Climate Change: Synthesis Report*, Washington, DC: World Bank.

About the authors

Rachel Berger is currently working independently on sustainability, climate change adaptation and community resilience issues. She worked for 11 years for Practical Action until 2012, latterly focusing on climate change adaptation, including community-based programmes and the development of international policy to support adaptation. She has written on a range of development subjects, most notably on CBA, co-authoring with Jonathan Ensor *Understanding Climate Change Adaptation: Lessons from Community-based Approaches*, published by Practical Action Publishing in 2009.

Jonathan Ensor is a lecturer at the Centre for Applied Human Rights, University of York, where he undertakes research, teaching and practice focused on the environment, development and human rights. He has written widely on CBA and the relationship between climate change and development practice, including *Uncertain Futures: Adapting Development to a Changing Climate*, published by Practical Action Publishing in 2011.

Part One
Thematic issues

CHAPTER 2

Power and politics in the governance of community-based adaptation

Julian S. Yates

This chapter presents a critical reflection on community-based adaptation, questioning the assumption that the community is a given and appropriate scale of governance and addressing the often neglected issue of multi-scalar flows of power and expertise. Policy prescriptions, governance decisions and the politics of adaptation are all produced at and travel across different scales, producing vulnerability and adaptation in the process. Extra-local economic, political and social forces are therefore as important as the nature of resources themselves. Thus, we need to distinguish between the act of adaptation and the informal rules or institutions that structure the ways in which technical expertise is transformed into adaptation. The aim here is to reveal these often hidden, multi-scalar political processes that determine governance contexts. The analysis is presented as a complement to, rather than a replacement for, quantitative policy instruments or normative frameworks for assessing adaptive capacity.

Keywords: power, politics, technology, community-based adaptation, governance, Nepal

Institutions and the governance of community-based adaptation

A growing body of work is paying attention to the ways in which the power and politics of decision making serve to structure community-based adaptation (CBA). This chapter addresses: how power flows through the technical interventions associated with adaptation to climate change (of non-governmental organizations (NGOs), for example); and how politics underpins the traditional notions of scale that often frame adaptation.[1] Climate-related policies and practices do not travel in unproblematic ways through governance scales and organizations to address 'hot spots' of vulnerability. We should not assume that communities reflect such hot spots or that the community scale is a pre-given and appropriate site for addressing vulnerability. Approaches that assume this fail to detect the power-laden elements of governing adaptation. These include: the ways in which adaptation can be used by powerful actors to achieve private political and economic gains; the unknown effects of adaptive practices; the inability to predict the ways in which adaptive capacity is transformed into adaptive action; and the ways in which behaviour, politics and

http://dx.doi.org/10.3362/9781780447902.002

risk intertwine (Adger and Barnett, 2009; Adger et al., 2003; Adger and Vincent, 2005; Anderson, 2010; Ensor and Berger, 2009b; Hulme, 2010; Moser, 2009).

There are limits, therefore, to how much we should focus our efforts on a restricted notion of CBA that casts communities as both the problem – as sites of vulnerability, lacking adaptive capacity – and the solution – as fixed sites open for technical intervention designed to put human ingenuity to work in overcoming the depredations of nature. In this chapter, I attempt to disrupt – through a sympathetic critique – any unwarranted comfort in the notion that CBA is somehow an inherently appropriate mechanism, scale or linear process. I draw attention to the limits of the perception that we can simply mould community institutions and organizations to provide a sturdy fulcrum beneath the levers of economic resources, technology, information and skills, and infrastructure – all of which can supposedly be put to the task of raising adaptive capacity (Moser, 2009). The point of the chapter is not to undermine CBA; on the contrary, the point is to help produce the analytical tools with which to uncover the processes that can make adaptation unequal, ineffective and unfair. These tools can help to reveal, and therefore prevent, the elite capture of development initiatives aimed at enhancing adaptive capacity and reducing vulnerability.

I begin by drawing attention to the ways in which the political production of vulnerability relies on technocratic expertise within global governance institutions and on a fixed notion of the community. I then argue that we must understand the community as a node in a network of diverse relations and flows of resources across scales. I follow this by presenting two potential ways of addressing these gaps in our understanding of CBA: 1) by recognizing the role of CBA within a politics of technical development; and 2) by understanding the dynamics of scalar politics in order to address the ways in which adaptive 'processes and institutionalized practices are themselves differentially scaled' (MacKinnon, 2011: 21). Towards the end of the chapter, I draw on experiences in Nepal and Peru to illustrate the value of a scalar political analysis of technical development. I conclude the chapter by bringing back a normative element, arguing in the process that we must think in terms of environmental justice.

Rethinking 'community'

Why is CBA being used as a lens through which to view scales of adaptation, or as a mechanism with which to carry out or promote particular adaptive practices? In this section, I point to the need to think critically about CBA in two related ways. First, we must recognize that communities are constructed according to scientific and political processes of identifying and then 'solving' vulnerability. Second, we must reflect on why we place a normative value on the notion of community and the process of CBA.

The political production of communities as hot spots of vulnerability

Current frameworks for addressing vulnerability to climate change attempt to measure the phenomenon as a function of exposure, sensitivity and adaptive capacity. This approach follows the conceptualization of vulnerability outlined by the Intergovernmental Panel on Climate Change (IPCC) (see Pachauri and Reisinger, 2007). It also reflects the demand within global governance frameworks (including the United Nations Framework Convention on Climate Change – UNFCCC) for rapidly produced, policy-relevant assessments that pinpoint locations for resource allocation. To complement the drafting of the National Adaptation Programme of Action (NAPA) of Nepal, for example, teams of experts from NGOs carried out local assessments of vulnerability according to a series of fixed and predetermined indicators. The local measurements were then aggregated into district and national assessments, with the aim of producing nation-wide maps of vulnerability hot spots. This approach reflects the general trend within frameworks being developed by NGOs and transnational institutions (such as the World Bank) to measure and assess vulnerability in order to steer development interventions (see, for example: Bizikova et al., 2009; Christian Aid, n.d.; Dazé et al., 2009; Hegglin and Huggel, 2008; Marshall et al., 2009; Panray et al., 2009; Pasteur, 2010; Regmi et al., 2010; Van den Berg and Feinstein, 2009; Wiggins, 2009).

Yet there are no measurable or generalizable '*a priori* factors, processes or functional relationships between exposure, sensitivity and adaptive capacity ... they are distinctive to particular places and times' (Smit and Wandel, 2006: 286). While relevant within today's regime of 'fast policy' – characterized by prescriptive forms of 'front-loaded advice and evaluation science' (Peck, 2011: 774) – the approaches above can compound uncertainty. Methods that aggregate vulnerability according to a fixed series of indicators cannot be relied upon to communicate complexities, and should not be used for the comparison of vulnerability as a context-specific problem (Barnett et al., 2008). Engle and Lemos's (2010) attempt at quantification, for example, fails to adhere to the complex web of qualitative relations that they argue characterizes the adaptive capacity of water governance institutions in Brazil. Many of these approaches are simply extensions of pre-existing frameworks (e.g. the sustainable livelihoods framework) for assessing and supporting normative processes of development according to the 'desirable states' of socio-ecological systems (see, for example: Nelson et al., 2007; Nelson et al., 2010). In addressing the vulnerability of rural Australian communities, Nelson et al. (2010) apply a static rural livelihoods analysis to quantify and subsequently aggregate adaptive capacity. Their assessment and subsequent recommendations for intervention revolve around the notion of 'asset substitution', which simply reflects a degree of 'absorbing capacity' (Ensor, 2011: 6) within a given (often fixed) livelihood system. Adaptive capacity, on the other hand, reflects the dynamic ability to reconfigure the livelihood system itself, thereby altering the very conditions of vulnerability.

Although some have tried to overcome these limitations through the use of participatory methods (see Fazey et al., 2010), frameworks such as Gupta et al.'s (2010) 'adaptive capacity wheel' subjectively measure the concept according to 22 criteria that are decided upon by Western scientists carrying predetermined yet unstated assumptions and belief systems (Hulme and Dessai, 2008). While attentive to the role of institutions and issues of equality, legitimacy, accountability, diversity and so on, Gupta and her colleagues ultimately suggest scoring these criteria on a scale from –2 to 2. They then argue that aggregation is not only possible but also desirable where it can facilitate comparison between places (thus ignoring Barnett et al.'s (2008) warnings).

Attempts to measure adaptive capacity and its institutional components in this way compound the uncertainty of the concept, since 'the very notion of governance indicators is problematic' (Adger and Vincent, 2005: 403). These attempts construct the community as a site of vulnerability and intervention, reducing complex social relations to hot spots on policy-relevant maps. Vulnerability, in this context, is *politically produced* as much as it is naturally given. Nonetheless, it is not my suggestion that such approaches should be abandoned. Rather, they must be recognized as (perhaps necessary) rapid, reductionist assessments that are conducted to guide the drafting and implementation of 'fast policy' (Peck, 2011: 773). To fully address the relational aspects of vulnerability and adaptive capacity, we need an approach that systematically teases out the ways in which local complexity co-determines vulnerability and adaptive capacity. Such an approach requires us to move beyond conceptions of the community as a fixed unit capable of being assessed and rebuilt according to prescriptive policy frameworks.

Communities as networks

The 'normative place' (Andolina et al., 2009: 80) of the community has emerged as a local site (or locality) that is supposedly good or appropriate for development practices and is a target of international development policies. These localities, however, are more than local: they are 'material, social, embodied, and networked resources for individual and communal agency' (Staeheli, 2008: 36). The community does not emerge 'in a political vacuum, but is produced through the exercise of power, as the outcome of negotiation, struggle, and compromise' (Perreault, 2003: 98). In this light, the community is not a given entity, a fixed scale, or inherently empowering. Neither, as Panelli and Welch (2005: 1593) have pointed out, is it 'comfortably uniform, complete, or blessed with consensus or agreement'. Rather than a unified product, the community is a process defined by contradictions and multiple perspectives, as well as elite control. It is important that we do not build an assumption of the community that wishes away the contested (disputed, challenged, conflicted) social and political relations upon which it rests, and that conceals the unevenness of vulnerability and adaptation to climate change.

Reconceiving the community as a network, rather than as a fixed location, complicates our understanding of CBA. Communities are based upon the intersection of diverse flows of relations and resources across scales – on a politics of scale and place. As Ensor (2011: 2) has previously pointed out, recognizing this element 'means that the quality of relationships, determined by characteristics such as power, culture, and gender, are drawn into the foreground so that interventions can identify the constraints on local decision making, looking across scales rather than at communities in isolation'. To understand the possibilities for knowledge generation and power sharing for fair and equal adaptation, we need to uncover this politics of place, scale and community. In what follows, I propose some ways forward in this regard.

Reframing the governance of community-based adaptation

In the previous section, I identified two under-addressed issues relating to the governance of community-based development. First, the notions of vulnerability, adaptation and community are all, in part, scientifically and politically produced. Second, communities should not be considered as fixed units that are at once the problem (i.e. they are vulnerable) and the solution (i.e. the assumed ideal site for technical solutions to vulnerability). To overcome this pitfall, I have suggested that we understand communities as networks of multiple contested relations (i.e. relations that are not given, homogeneous, nor unproblematic to the community). These relations flow across and co-produce the very scales at which climate change and adaptation governance practices work. In this section, I suggest that we can systematically address the above issues by viewing CBA through the lenses of techno-politics and scalar politics. These approaches help us to come to terms with the ways in which power flows through social networks to determine what knowledge is used by whom, for what kind of adaptive practices, and for whose benefit.

The politics of technical development

Recognizing technical development as a political issue, or as what Mitchell (2002) refers to as techno-politics, reconceives notions such as knowledge transfer and technical assistance. This approach draws attention away from unproblematic flows of information and knowledge, and focuses instead on a contested politics of technical development (that is, how technical development is determined by fraught political processes). In this light, adaptation is shaped by a politics of technology and development. This concept gives rise to three implications for the implementation of technical projects to enhance adaptive capacity. First, given the historically constituted discourses and practices (such as neo-colonialism) through which development programmes arise (de la Cadena, 2010), technical projects cannot be separated from political interventions. NGO programmes are implicated in this politics, as their interventions are place-based and contextualized, and yet simultaneously connected to broader flows of knowledge, resources, ideas, values and power (Bebbington, 2004). Both government and non-government interventions designed to enhance adaptive capacity are tied to a broader politics of knowledge production and technical expertise.

Second, as technical interventions, NGO projects and programmes are tied to the institutions of science and technology, which shape what activities occur in a particular place (Mitchell, 2002). In this context, we must make the distinction between the *act* of adaptation and the informal rules or institutions that *structure* adaptation. In Mitchell's account, technical development projects at a variety of scales begin to construct what is acceptable knowledge and science, as well as to produce concentrations of power and control. Not only are technical interventions inherently political, they also have political *effects*. However, technical expertise is not introduced from the outside in a linear fashion; it is a reorganization of new and existing knowledge and power, and is 'always somewhat overrun by the unintended' (Ferguson, 1994; Mitchell, 2002: 42–3). Interventions that attempt to build adaptive capacity do not yield predictable and unproblematic results; community adaptive capacity is not enhanced in unison, as a whole capable of overcoming the new vulnerabilities introduced by climate change. There are always unintended consequences that benefit some people at the expense of others. Knowledge, expertise and the discourses of Western science intersect with local norms and customs to institutionalize certain practices and structure the ways in which adaptive projects are implemented.

Third, framed as an issue of technical development, the process of adaptation is inherently political and technological; adaptation is shaped by a politics of technology and development. Adaptation relies, in part, on experts and powerful political players. While the existence and availability of knowledge and information are essential, they are never enough on their own; knowledge is not simply a means for overcoming uncertainty or unequal power relations, and questions relating to the use of information in decision-making arenas are essential (Adger et al., 2009; Moser, 2009). Useful knowledge comes into being through particular social and political orderings and under particular institutional arrangements that are underpinned by power relations between political actors (Hulme, 2010).

We must turn our attention, then, to the hybrid nature of knowledge, and to the ways in which scientific knowledge of climate change interacts with local customs, as well as with the power relations that transform such knowledge into adaptive practice. This realization means that CBA becomes 'about the issue of power and agency as a question, not an answer known in advance' (Mitchell, 2002: 53). Raising the question of power and tracking the production, utilization and effects of human agency become fundamental to processes of uncovering how CBA functions. The concept of techno-politics helps by seeking out the connections and interactions that portray adaptive actions as the human capacity to overcome vulnerability to nature (Mitchell, 2002). This perspective, and the elements outlined in Box 2.1, help us to understand community vulnerability as part of a political process and to reframe CBA as a form of techno-political development intervention.

Box 2.1 Key elements of techno-politics

Technology is political: technological interventions are not neutral. Technologies constitute, embody and/or enact political goals. Technical approaches to adaptation always entail *unintended outcomes* as they come into contact with particular human and non-human contexts.

Similarly, *expertise* is not neutral. Rather than simply reflecting social relations, experts (those with technical knowledge) from different contexts will inevitably shape and influence relationships.

Hybrid agencies/institutions emerge where external expertise meets local knowledge and practice. Recognizing how informal rules *structure* adaptation exposes the techno-politics of adaptation.

Scalar politics

Emerging from a fixed notion of community is the assumption that adaptive practices occur at a *particular* scale (often the local or community), which can be conveniently nested within others (regional, global, etc.) (see, for example: Adger, 2001; Sikor et al., 2010). Terms such as 'cross-scale' have often been used as a synonym for horizontal social networks *within* a particular scale, such as the local (e.g. Osbahr et al., 2008). Research into the scalar elements of adaptation must move beyond these fixed approaches to address processes within, between and across scales, while seeking not simply to combine vertical (hierarchical) and horizontal perspectives (Reed and Bruyneel, 2010). In particular, attention has turned to the notion of scale-jumping or scale-crossing, which refers to the ability of social groups and organizations to shift between levels of activity in order to influence practical outcomes in policy (Benson, 2010; Triscritti, 2013). While such cross-level interactions are supposedly 'widespread and ... have important consequences for efforts to govern human/environment relationships' (Young, 2006: 27), these concepts treat scale as an 'already partitioned geography' distinct from social practices (MacKinnon, 2011: 24). The assumption exists that scales of decision making are given precursors of social activity, rather than the product of social and political practices.

In contrast, paying attention to a scalar politics can help to unravel the 'strategic deployment of scale' (MacKinnon, 2011: 29) by various actors seeking to affect adaptation decisions; scale emerges as a dimension of political activity. Scalar political analysis is a useful complement to (rather than replacement for) other modes of analysis, such as quantitative policy instruments, normative frameworks or heuristic devices deployed *at particular scales*. Scalar political analysis helps to reveal what may sometimes go undetected: a complex and unequal flow of adaptation decisions and actions within and across scales of governance; and the rearrangement of existing scales of decision making in order to serve adaptation needs.

In the next section, I illustrate the utility of scalar political analysis by drawing explicitly on MacKinnon's four broad principles of scalar politics: the scalar aspects and repercussions of political projects and initiatives;

the strategic deployment of scale; the continuing influence and effect of pre-existing scalar structures; and the creation of new scalar arrangements. These four principles are not wholly distinct but rather they overlap, as social practices and processes have multiple scalar effects.[2] I draw out this framework to illustrate its value – in conjunction with the lens of techno-politics – in helping us to reinterpret contexts of vulnerability as well as adaptive capacity and action.

The scalar techno-politics of governing adaptation

In the previous section, I pointed to the analytical approaches of techno-politics and scalar politics as ways of systematically coping with the limits of community-focused approaches (outlined earlier in the chapter). Understanding techno-politics helps to reveal the ways in which technical expertise and politics intertwine to produce community vulnerability as well as hybrid technical solutions. These solutions are hybrid because they rely on both flows of 'external' expertise and local or 'internal' knowledge. They occur at the points where multiple, diverse flows of knowledge, resources and power combine to determine material outcomes. Scalar political analysis helps in understanding how hybrid solutions emerge, as it moves us beyond the fixed notion of the community and points to the multi-scalar institutional processes that structure adaptation. A scalar political analysis therefore helps us to see the ways in which the techno-politics of adaptation is produced and reproduced across scales of decision making. In this section, I illustrate the usefulness of understanding the scalar techno-politics of adaptation in order to uncover the elements of CBA that often go undetected.

The scalar aspects and repercussions of political projects and initiatives

There are frictions involved in the process of identifying and targeting sites of vulnerability, as networks of decision making collide in projects of CBA. Importantly, NGOs and their projects are now heavily implicated in shaping scalar repercussions, as new forms of connectivity rework places and livelihoods (Bebbington, 2004). In Nepal, for example, flows of expertise and technological interventions have had the unintended consequence of redefining scales of decision making. In the district of Chitwan, tensions exist between the technical interventions of NGOs, the management strategies of Chitwan National Park, the role of community forestry programmes, and the livelihood needs of local residents. The erection of an electric fence, as part of a so-called 'multi-stakeholder initiative' to prevent wildlife intrusion, did not simply reduce the intrusion of rhinos into villages as intended. The demand for grasses (used by local residents as animal fodder, insulation and fuel) has outstripped supply in a community forest increasingly frequented by the rhinos in search of the same grasses. With the rhinos also squeezed on the other side by the reduced availability of grasses in Chitwan National Park, the community forest is struggling to supply them all.

In response, villagers have travelled in the opposite direction, as they have begun foraging illegally for grasses in the National Park by crossing the dangerously erratic Rapti River, thereby blurring the boundaries between the community forest and the park. Politically, the Chitwan National Park has been turning a blind eye to both rhino intrusion and foraging in the park, while NGO influence diminished after the erection of the electric fence. It has been left to the community forest user groups to determine how to manage the conflict over forest resources, as well as navigate the material and political–institutional boundaries of the National Park, the community forest, the village, and the diminishing role of local government (the village development committee – VDC). The local influence of the transnational NGOs therefore goes beyond the prevention of rhinos entering villages; decision making has shifted from the National Park and VDC to community groups and their networked relations with the often sporadic interventions of those NGOs.

The effects of installing the electric fence illustrate that the networks of expertise embodied in NGO interventions have had unintended impacts on scales of decision making and have not always produced successful material results. This point is also illustrated in Nepal by the failure of an NGO-instigated bamboo plantation, designed to rejuvenate riparian vegetation, reduce flood frequencies and enhance socioeconomic livelihoods through the sale of non-timber forest products. The failure of the plantation to quell overland water flow and restore reservoir levels has meant that drought conditions have not been mitigated. Local farmers have placed the blame for the drying Baulaha River on the nearby upstream community, arguing that its uninformed practices have tampered with watershed dynamics. Rather than assert their own solutions, downstream community members have been demanding more NGO interventions (such as requesting that NGOs provide improved seed varieties). These community-wide demands, presented formally by local key players, position the community as technologically rather than materially vulnerable, and therefore in need of insertion into broader networks of expertise and technical innovation. Ironically, upstream community members are aware that yam cultivation is a viable alternative in times of drought, yet they remain unable to transport the yams to nearby markets. The upstream community is, therefore, confident in its traditional knowledge of yam cultivation and positions itself as physically isolated due to failures in local infrastructure development, rather than politically isolated from global flows of knowledge and expertise through NGOs. To understand the ways in which adaptive practices take shape in local contexts, we need to be attentive to the ways in which communities position themselves within the politics of technical development.

The techno-political interventions of NGOs also serve to instil particularly scaled responses to the challenges of climate change. The prevalence of disaster risk reduction (DRR) perspectives, for example, often reflects a 'fast policy' approach that places emphasis on local solutions to more-than-local problems, and is technocratically produced and reproduced by actors in multilateral agencies (Peck and Theodore, 2010). The current response environment in Chitwan is characteristic of these combined structural pressures, as decision makers in local government have emphasized short-term, local technocratic options (e.g. temporary shelters) over issues such as education facilities and increased security of land tenure arrangements. Techno-political approaches to DRR reflect a paradox of adaptation interventions: supported by global policy prescriptions in response to the global challenge of climate change, such DRR perspectives continue to assert the local as an appropriate scale of problem solving more-than-local problems. As Castree (2001: 11) has pointed out, the 'technofix policies so beloved of conventional hazard managers [have often] failed to address the deeper issues of why certain social groups ... are more vulnerable to hazards in the first place'.

The strategic deployment of scale

According to MacKinnon (2011: 29), 'scalar politics focuses attention on the strategic deployment of scale by various actors, organizations, and movements'. Powerful political actors can manipulate a network of multi-scalar alliances in order to construct and preserve the local conditions that produce their power and material well-being (Cox, 1998). In the context of adaptation, political affiliation across scales is used by elites to reinforce their already strong grip on local resource allocation. This approach adds a scalar dimension to existing approaches that explore the role of key players, who are 'embedded in particular institutional, normative, and political contexts – at the centre of governance' (Moser, 2009: 315). Political agency is not distributed equally and it is perhaps unsurprising that strategies for adaptation are often formulated by dominant and powerful elites. This form of elite capture carries an important scalar element: in order to control local spaces of adaptation, local elites must jump and bend scales in order to participate in decision-making processes at broader scales of governance and stretch the influence of local institutions in determining and implementing adaptation projects. The term 'scale bending' was coined by Neil Smith (2004) to refer to the processes of and struggles over stretching, fragmenting and reconstructing previously established and accepted political scales. Scale bending challenges existing scalar arrangements (and the identities tied to those arrangements) by undermining the ways in which particular social relations are tied to certain scales. The point is that challenges to existing scales of decision making do not simply occur at one scale but result from interactions at several scales simultaneously.

In Nepal, for example, key players have been able to use their political ties to secure resources on the basis of their position within a 'vulnerable community'. One key player in particular exerted influence over the (supposedly democratic) decision-making processes within the community forestry programme and exploited personal affiliation within multi-scalar political networks in order to gain direct access to resources and project approval at the district level and drive forward a particular agenda for adaptation. The slow and bureaucratic process of seeking resources and project support through the VDC was circumvented; the scale of decision making was bent around the VDC, as the key player tied the community forestry programme to decision making at the district level and to projects implemented by transnational NGOs. The key player acted as an 'expert' in Mitchell's (2002) sense of becoming a spokesperson for a particular and partial view of how development decisions should be made.

The power of these key players is initially derived from horizontal connections between a variety of government and non-government institutions and organizations at the local level. This position enables them to 'pull down' resources – directly to the local level – from other (vertical) scales of governance, including from international NGOs operating within the (global) framework of the UNFCCC and the (global/national) framework of NAPAs. Thus, we begin to see the production of community vulnerability in a new light: communities are not simply constructed as vulnerable from above by broad governance mechanisms such as the NAPAs. Rather, local key players are able to tap into these evolving governance frameworks to recast how communities are perceived and to redirect resources. Key players are not simply 'jumping' scales (which implies that participation can occur only at distinct scales) but are bending or blurring the very distinction between scales. While they appear to be active at just one scale, they are in fact using their power and influence to tap into the proliferating resources available at the transnational, national and district scales for supporting local adaptive practices. Multiple, scaled processes of governance therefore coalesce, as technical NGO interventions 'trickle down' and local interests are 'pushed up' by key players. In this context, assessments of vulnerability to climate change – which inform the drafting of NAPAs – are not so much about the aggregate scores generated by many technocratic instruments, but more about a scalar politics that defines power relations and decision-making processes in many different governance mechanisms. The eventual content of both the local and the national adaptation plans of action is likely to be as much a product of how actors are able to influence decision-making structures as it is about material vulnerabilities.

The continuing influence and effect of pre-existing scalar structures

Pre-existing scalar structures continue to have an influence and effect (MacKinnon, 2011). There is, then, a materiality to scale that emerges from both inherited institutional structures and the 'more-than-human' (Dowling, 2010: 492) biophysical processes and contexts sometimes referred to as natural hazards. Importantly, material conditions do not simply determine scales, but are also affected by the scalar aspects of social practices. In the case of the wildlife fence in Chitwan, for example, the material (more-than-human) elements – the rhinos, the erratic Rapti River, vegetation properties – and the existing scalar institutions of the community forest and the National Park all play a role in the ways in which emerging social practices reconfigure scalar arrangements.

The increasingly pivotal role of water resources in Nepal provides a further illustration. Receding glaciers and reduced snowmelt volumes combine with reduced rainfall in the low-lying plains to exacerbate the intensity and duration of drought conditions. Disputes over water access have become a central concern of climate change adaptation debates in the country, disputes that reflect the tensions between existing institutional scales and those required for adaptive governance. Despite their inclusion in the formal National Water Plan, the current weak status of water user associations in Nepal means that they are poorly equipped to rise to the challenge of spanning the particular interests of different water users, including individuals, private sector actors and communities (Yates, 2012). This lack of effectiveness reflects the fact that a formal 're-scaling' of governance to the local level does not necessarily imply greater empowerment for local actors, as different levels of government continue to be important (Norman and Bakker, 2009), especially when exploited by the strong network linkages of key players.

Where communities lack strong networks of political affiliation, the formal 're-scaling' of governance mechanisms to the local level can often be symbolic, with pre-existing mechanisms and scales continuing to dominate decision making. For members of the yam-producing upstream community introduced above, participation in decision-making scenarios is often symbolic and does not affect the outcome of decisions or the development plans that emerge. Although these residents are engaged in local watershed management practices (such as the construction of irrigation channels, afforestation to reduce overland flow, etc.), their lack of capacity to affect scaled networks of political decision making means that they have little say over processes occurring downstream. Without the ability to circumvent the existing institutional architecture in favour of new forms of scalar representation (such as drawing resources directly from the district development committee (DDC), or constructing a new scale of watershed management), the upstream residents struggle to engage in a form of scalar politics that can positively and productively affect change.

Elsewhere, however, existing institutional mechanisms are being revived and transformed in order to help build adaptive capacity. In Peru, the two institutions of *ayllu* (a traditional collective arrangement) and *kamayoq* (a mobile mechanism for knowledge exchange) are becoming increasingly important. The *ayllu* incorporates communal and reciprocal arrangements for labour and information exchange, and has provided the building blocks for the emergence of community committees for risk management – community-based organizations that focus on reducing vulnerability through information dissemination and improving preparedness for extreme events (Ensor and Berger, 2009a). To mobilize the resources necessary for reacting to sudden events, these committees co-ordinate activities along with communities and regional and local government. What was a localized institution for organizing social life and livelihoods has become an important mechanism within a multi-scalar structure for enhancing adaptive capacity that builds on, rather than overrides, the strengths of existing customs and norms.

Likewise, the importance of *kamayoq* within a multi-scalar system of mobile knowledge exchange has prompted the resurgence of the phenomenon within local development and adaptation projects. While the term descends from the Quechua verb '*kamay*' – to be in charge of or over (Adelaar and Muysken, 2004: 494; de la Vega, 1609: 97; Gonzalez Holguin, 1952) – contemporary usage within development projects refers to 'specialists' within particular spheres of production and/or social-cultural practice (de la Torre Postigo, 2004; van Immerzeel, 2006). *Kamayoq* extend adaptation policy and practice to its furthest reaches in the remote Peruvian Andes. Importantly, the training and certification of *kamayoq* has become a core component of development projects led by NGOs and government development agencies. The recent formal certification by government has also served to institutionalize the *kamayoq* system, lending legitimacy to their role as extension agents but also incorporating them within flows of techno-political development. The training programme links the *kamayoq* to the flows of knowledge and expertise that are carried by international NGOs from global governance frameworks to the particular places that are deemed vulnerable. The *kamayoq* system therefore helps to extend the techno-politics of development beyond the immediate reaches of the NGO. The *kamayoq* system may be read in Mitchell's sense as a hybrid that combines external forms of expertise with existing knowledge and practices that operate according to scales rarely identified by international climate change institutions. To fully understand the effect of the *kamayoq* system on adaptive capacity and adaptive action, we must develop a scalar political analysis that locates it within this network of development techniques – a network that spans scales of decision making, cultures of practice and systems of knowledge production.

The creation of new scalar arrangements

Finally, there is a continuously shifting relation between scale and political process; while inherited institutional structures cannot be escaped, they are also continuously being reshaped. The UNFCCC and the IPCC are clear examples of new global governance structures that have emerged in an attempt to match the global nature and effects of climate change. These mechanisms, however, remain constrained by territorial and economic logics of nation states (whereby they must safeguard national interests and maintain competitiveness) and reinforce the framing of climate change vulnerability and adaptive capacity as a national issue. NAPAs fix adaptation to the national scale by placing a focus on cost–benefit analyses of climate-related impacts and potential adaptive action. They also focus on measurable variables such as economic well-being, rather than on less quantifiable elements such as cultural and institutional regimes, long-term processes and attachments to place (Ribot, 2011).

Nonetheless, feeding into national and international governance mechanisms is a host of new institutional arrangements, which emerged specifically from climate change-related action and which requires greater attention within work on scales of governance. The examples from Peru explored in the previous section illustrate how existing institutions can bring these new scalar arrangements to CBA. In Nepal, however, entirely new institutional mechanisms have emerged specifically to tackle climate change issues. Nepal's NAPA will be implemented from a base of local adaptation plans of action, which promote a bottom-up process of decision making for adaptive action. While VDCs may be involved in the process, they are no longer the principal agency responsible for governing development projects. Rather, it is the newly formed district co-ordination committees – consisting of multiple district-level government offices, NGOs, community organizations, private enterprises and local service delivery agents – that will govern climate adaptation projects. These new mechanisms may reflect a form of polycentric governance similar to that called for by Elinor Ostrom (2010), but as they evolve, close attention must be paid to the ways in which decisions are made and how adaptive actions are put in place. That is: who is deciding, for whom, what kinds of adaptive projects should address what kinds of climate-related issues? The mere fact that new governance mechanisms have emerged does not mean that we should assume that they will be any more effective or representative.

It is not just formal processes of reorganizing governance structures that have produced new scalar arrangements in Nepal. Disaster management committees (DMCs), for example, have emerged from networked interactions between various scales of government, international and domestic NGOs, local organizations and local residents. Decision-making procedures within the committees supposedly follow the model established by the community forestry programme, which is based on an elected assembly of local representatives. However, the DMCs also reflect (though they may conceal) multiple scaled forms of representation and decision making. For example, while DMCs appear to offer a democratic and local mode of governance, the committees in fact crystallize multiple scales of governance, knowledge and expertise into a new institutional form. In this context, key players and institutions that operate across scales (such as NGOs and their representatives) possess a functional influence over the local everyday practices and decision-making procedures of the committee. Through this technocratic influence, the DMCs take on the characteristics of Mitchell's (2002) hybrid agencies and begin to portray short-term actions as forms of human expertise capable of overcoming vulnerability to nature. In particular, the heavy emphasis on DRR technologies, such as temporary shelters, prioritizes the human capability of coping with natural hazards in the short term, thereby overshadowing long-term transformation and adaptation.

The DMCs reflect a transformation of traditional scalar hierarchies and existing institutional arrangements as they bring various institutions both horizontally (e.g. the VDC, community organizations, user groups and committees) and vertically (the DDC and NGOs) into a tighter network for the governance of diffuse property rights within disaster management. They are also hybrid institutions in Andolina et al.'s (2009) sense, in that they have become important in stimulating knowledge, resources and personnel to move across scales of decision making that were previously socially assumed to be fixed. Although these hybrid institutions are constituted by multi-scalar links, they also reinforce local adaptation agendas. While appearing to operate like and descend from the community forestry model, DMCs are a local mutation of policies and practices that operate within a contemporary context of multi-scalar climate change governance.

Conclusion: addressing power

In this chapter I have introduced just some of the ways in which we need to reflect more systematically on the notion of CBA. In particular, we need to think about *why* a community-based approach is deemed appropriate and necessary, rather than simply assume that it is at the community scale where we can see and tackle vulnerability most effectively. I have argued that we must be attentive to more nuanced understandings of communities as networks that are structured by unequal power relations and unequal access to knowledge, resources and decision-making structures. To understand the ways in which social groups are vulnerable and how adaptive practices and processes play out on the ground, we must come to terms with this power and politics of CBA. In an attempt to do so, I illustrated the value of reading CBA through the lens of scalar techno-politics. I do not suggest that such a reading supersedes other modes of analysis, such as Nelson et al.'s (2010) use of a livelihoods framework to make sense of adaptive capacity. Rather, the kinds of analysis presented here are a necessary complement to other approaches in order to address the unequal and unfair nature of CBA.

Finally, therefore, it is necessary to return to the question of how we develop a normative approach to CBA in light of the above discussion. The notions of techno-politics and scalar politics reveal how networks of power and influence flow across and reshape the scalar arrangements and institutions that structure adaptation. The effects are unequal and can lead to the elite capture of adaptive action. How can we address this problem? For Ensor (2011: 54), the answer lies in rights-based approaches, which he argues 'bring power to the foreground of development practice'. However, without addressing precisely what will be ascribed the status of a 'right' within CBA, it remains unclear precisely how a rights-based approach would function. Even where a particular public good can be identified, such as water, there remain limits to the ability of rights-based approaches to overcome the effects of unequal power relations and to address complex, collective governance issues (Bakker, 2007; 2010). Lacking such a clearly identifiable public good, the lessons presented by Ensor (2011) are not just about rights per se, but more about justice – about demands to participate substantively in decision making, to hold the state accountable, and to use the law to challenge rights violations.

What we need, then, is an environmental *justice* perspective that builds on these lessons of rights-based approaches, but moves beyond them to tackle the real and unequal causes of vulnerability. If being poor and vulnerable is 'not what people are, but what people are made' (Mander, quoted in Ensor, 2011: 54), then development and vulnerability reduction are not about the right not to be poor and vulnerable, but are rather about the *injustice* of being made poor and vulnerable. The techno-politics that I have explored here plays a part in this injustice, but it can also contribute to potential solutions. We need a perspective that builds an understanding of the varied causes and scales of environmental injustice and reveals the diversity and place specificity of definitions and articulations of environmental justice (Holifield et al., 2009). The perspectives presented in this book that are written *from*, rather than *for*, the community perspective may help to shed light on the diverse political processes through which the varied scales of environmental vulnerability and injustice can be addressed. These perspectives are required in order to highlight hybrid spaces of action, which emerge from innovative cases of interaction – cases of shaping development projects and processes *from*, rather than *for*, local knowledge and practices. By placing emphasis on marginalized forms of knowledge, these spaces may directly tackle unequal networks of power to create hybrid social practices that can contribute to vulnerability reduction among the politically marginalized and the most vulnerable. These spaces highlight the particular ways in which hybrid social practices have just or unjust local effects, thereby helping communities of practice (which include local residents, NGOs, government bodies, etc.) to make more informed choices about who benefits from, and what conditions are challenged by, particular kinds of adaptive action.

Notes

1 Traditional conceptions of scale often distinguish between 'vertical' and 'horizontal'. The former refers to the relations among 'nested territorially defined political entities' (Leitner, 2004: 237) – in the case of Nepal, for example, this may refer to the nesting of the VDC within the DDC, the DDC within development regions, and these regions within the national scale. 'Horizontal' refers to links across geographic space – to connections that link together 'institutions and actors across the boundaries that mark the traditional geography of political power' (Leitner, 2004: 236–7).

2 Given the aims of this collection and the space constraints involved, a full theoretical discussion of the concept of scalar politics is not possible; for a discussion in the context of adaptation, see Yates (2012).

References

Adelaar, W. and Muysken, P. (2004) *The Languages of the Andes*, Cambridge: Cambridge University Press.

Adger, W.N. (2001) 'Scales of governance and environmental justice for adaptation and mitigation of climate change', *Journal of International Development* 13: 921–31 <http://dx.doi.org/10.1002/jid.833>.

Adger, W.N. and Barnett, J. (2009) 'Four reasons for concern about adaptation to climate change', *Environment and Planning A* 41: 2800–5 <www.envplan.com/abstract.cgi?id=a42244>.

Adger, W.N., Dessai, S., Goulden, M. et al. (2009) 'Are there social limits to adaptation to climate change?', *Climatic Change* 93: 335–54 <http://dx.doi.org/10.1007/s10584-008-9520-z>.

Adger, W.N., Huq, S., Brown, K., Conway, D., and Hulme, M. (2003) 'Adaptation to climate change in the developing world', *Progress in Development Studies* 3: 179–95 <http://dx.doi.org/10.1191/1464993403ps060oa>.

Adger, W.N. and Vincent, K. (2005) 'Uncertainty in adaptive capacity', *Comptes Rendus Geosciences* 337: 399–410 <http://dx.doi.org/10.1016/j.crte.2004.11.004>.

Anderson, B. (2010) 'Preemption, precaution, preparedness: anticipatory action and future geographies', *Progress in Human Geography* 34: 777–98 <http://dx.doi.org/10.1177/0309132510362600>.

Andolina, R., Laurie, N. and Radcliffe, S.A. (2009) *Indigenous Development in the Andes: Culture, Power, and Transnationalism,* Durham, NC: Duke University Press.

Bakker, K. (2007) 'The "commons" versus the "commodity": alter-globalization, anti-privatization and the human right to water in the global south', *Antipode* 39: 430–55 <http://dx.doi.org/10.1111/j.1467-8330.2007.00534.x>.

Bakker, K. (2010) *Privatizing Water: Governance Failure and the World's Water Crisis,* Ithaca, NY: Cornell University Press.

Barnett, J., Lambert, S. and Fry, I. (2008) 'The hazards of indicators: insights from the Environmental Vulnerability Index', *Annals of the Association of American Geographers* 98: 102–19 <http://dx.doi.org/10.1080/00045600701734315>.

Bebbington, A.J. (2004) 'NGOs and uneven development: geographies of development intervention', *Progress in Human Geography* 28: 725–45 <http://dx.doi.org/10.1191/0309132504ph516oa>.

Benson, M.H. (2010) 'Regional initiatives: scaling the climate response and responding to conceptions of scale', *Annals of the Association of American Geographers* 100: 1025–35 <http://dx.doi.org/10.1080/00045608.2010.497317>.

Bizikova, L., Dickinson, T. and Pintér, L. (2009) 'Participatory scenario development for translating impacts of climate change into adaptations', in Ashley, H., Kenton, N. and Milligan, A. (eds), *Participatory Learning and Action No. 60: Community-based Adaptation to Climate Change,* London: International Institute for Environment and Development (IIED), pp. 167–72.

Castree, N. (2001) 'Socializing nature', in Castree, N. and Braun, B. (eds), *Social Nature: Theory, Practice and Politics,* Oxford: Blackwell, pp. 1–21.

Christian Aid (no date) *Adaptation Toolkit: Integrating Adaptation to Climate Change into Secure Livelihoods,* London: Christian Aid.

Cox, K.R. (1998) 'Spaces of dependence, spaces of engagement and the politics of scale, or: looking for local politics', *Political Geography* 17: 1–23.

Dazé, A., Ambrose, K. and Ehrhart, C. (2009) *Climate Vulnerability and Capacity Analysis,* London: CARE International.

de la Cadena, M. (2010) 'Indigenous cosmopolitics in the Andes: conceptual reflections beyond "politics"', *Cultural Anthropology* 25: 334–70 <http://dx.doi.org/10.1111/j.1548-1360.2010.01061.x>.

de la Torre Postigo, C. (2004) *Kamayoq: Promotores Campesinos de Innovaciones Tecnológicas,* Lima: Soluciones Prácticas.

de la Vega, G. (1609) *Comentarios Reales,* Lisbon: Pedro Crasbeck.

Dowling, R. (2010) 'Geographies of identity: climate change, governmentality and activism', *Progress in Human Geography* 34: 488–95 <http://dx.doi.org/10.1177/0309132509348427>.

Engle, N.L. and Lemos, M.C. (2010) 'Unpacking governance: building adaptive capacity to climate change of river basins in Brazil', *Global Environmental Change* 20: 4–13 <http://dx.doi.org/10.1016/j.gloenvcha.2009.07.001>.

Ensor, J. (2011) *Uncertain Futures: Adapting Development to a Changing Climate,* Rugby: Practical Action Publishing.

Ensor, J. and Berger, R. (2009a) 'Community-based adaptation and culture in theory and practice', in Adger, W.N., Lorenzoni, I. and O'Brien, K. (eds), *Adapting to Climate Change: Thresholds, Values, Governance,* Cambridge: University of Cambridge Press, pp. 227–39.

Ensor, J. and Berger, R. (2009b) *Understanding Climate Change Adaptation: Lessons from Community-based Approaches,* Rugby: Practical Action Publishing <http://dx.doi.org/10.3362/9781780440415>.

Fazey, I., Kesby, M., Evely, A., et al. (2010) 'A three-tiered approach to participatory vulnerability assessment in the Solomon Islands', *Global Environmental Change* 20: 713–28 <http://dx.doi.org/10.1016/j.gloenvcha.2010.04.011>.

Ferguson, J. (1994) *Anti-Politics Machine: Development, Depoliticization, and Bureaucratic Power in Lesotho,* Minneapolis, MN: University of Minnesota Press.

Gonzalez Holguin, D. (1952) *Vocabulario de la Lengua General de Todo el Perú llamada Lengua Qquichua o del Inca,* Lima: Universidad Nacional Mayor de San Marcos.

Gupta, J., Termeer, C., Klostermann, J. et al. (2010) 'The Adaptive Capacity Wheel: a method to assess the inherent characteristics of institutions to enable the adaptive capacity of society', *Environmental Science & Policy* 13: 459–71 <http://dx.doi.org/10.1016/j.envsci.2010.05.006>.

Hegglin, E. and Huggel, C. (2008) 'An integrated assessment of vulnerability to glacial hazards : a case study in the Cordillera Blanca, Peru', *Mountain Research and Development* 28: 299–309 <http://dx.doi.org/10.1659/mrd.0976>.

Holifield, R., Porter, M. and Walker, G. (2009) 'Introduction: spaces of environmental justice: frameworks for critical engagement', *Antipode* 41: 591–612 <http://dx.doi.org/10.1111/j.1467-8330.2009.00690.x>.

Hulme, M. (2010) 'Problems with making and governing global kinds of knowledge', *Global Environmental Change* 20: 558–64 <http://dx.doi.org/10.1016/j.gloenvcha.2010.07.005>.

Hulme, M. and Dessai, S. (2008) 'Negotiating future climates for public policy: a critical assessment of the development of climate scenarios for the UK', *Environmental Science & Policy* 11: 54–70 <http://dx.doi.org/10.1016/j.envsci.2007.09.003>.

Leitner, H. (2004) 'The politics of scale and networks of spatial connectivity: transnational interurban networks and the rescaling of political governance in Europe', in Sheppard, E. and McMaster, R. (eds), *Scale and Geographic Inquiry: Nature, Society, and Method,* Oxford: Wiley-Blackwell, pp. 237–55.

MacKinnon, D. (2011) 'Reconstructing scale: towards a new scalar politics', *Progress in Human Geography* 35: 21–36 <http://dx.doi.org/10.1177/0309132510367841>.

Marshall, N.A., Marshall, P.A., Tamelander, J. et al. (2009) 'A framework for social adaptation to climate change: sustaining tropical coastal communities and industries', Gland, Switzerland: International Union for Conservation of Nature (IUCN).

Mitchell, T. (2002) *Rule of Experts: Egypt, Techno-Politics, Modernity,* Berkeley, CA: University of California Press.

Moser, S.C. (2009) 'Whether our levers are long enough and the fulcrum strong? Exploring the soft underbelly of adaptation decisions and actions', in Adger, W.N., Lorenzoni, I. and O'Brien, K. (eds), *Adapting to Climate Change: Thresholds, Values, Governance,* Cambridge: University of Cambridge Press, pp. 313–34.

Nelson, D.R., Adger, W.N. and Brown, K. (2007) 'Adaptation to environmental change: contributions of a resilience framework', *Annual Review of Environment and Resources* 32(1): 395–419 <http://dx.doi.org/10.1146/annurev.energy.32.051807.090348>.

Nelson, R., Kokic, P., Crimp, S. et al. (2010) 'The vulnerability of Australian rural communities to climate variability and change: Part II – integrating impacts with adaptive capacity', *Environmental Science & Policy* 13: 18–27 <http://dx.doi.org/10.1016/j.envsci.2009.09.007>.

Norman, E.S. and Bakker, K. (2009) 'Transgressing scales: water governance across the Canada–U.S. borderland', *Annals of the Association of American Geographers* 99: 99–117 <http://dx.doi.org/10.1080/00045600802317218>.

Osbahr, H., Twyman, C., Adger, W.N., and Thomas, D.S.G. (2008) 'Effective livelihood adaptation to climate change disturbance: scale dimensions of practice in Mozambique', *Geoforum* 39(6): 1951–64 <http://dx.doi.org/10.1016/j.geoforum.2008.07.010>.

Ostrom, E. (2010) 'Polycentric systems for coping with collective action and global environmental change', *Global Environmental Change* 20: 550–57 <http://dx.doi.org/10.1016/j.gloenvcha.2010.07.004>.

Pachauri, R.K. and Reisinger, A. (2007) *Climate Change 2007: Synthesis Report. Contribution of Working Groups I, II and III to the Fourth Assessment Report of the Intergovernmental Panel on Climate Change,* Geneva: IPCC.

Panelli, R. and Welch, R. (2005) 'Why community? Reading difference and singularity with community', *Environment and Planning A* 37: 1589–611 <http://dx.doi.org/10.1068/a37257>.

Panray, K.B., Noyensing, G. and Reddi, K.M. (2009) 'Vulnerability assessment as a tool to build resilience among the coastal community of Mauritius', in Van den Berg, R.D. and Feinstein, O. (eds), *Evaluating Climate Change and Development,* Washington, DC: World Bank, pp. 361–78.

Pasteur, K. (2010) *Integrating Approaches: Sustainable Livelihoods, Disaster Risk Reduction and Climate Change Adaptation,* Rugby: Practical Action Publishing.

Peck, J. (2011) 'Geographies of policy: from transfer-diffusion to mobility-mutation', *Progress in Human Geography* 35: 773–97 <http://dx.doi.org/10.1177/0309132510394010>.

Peck, J. and Theodore, N. (2010) 'Recombinant workfare, across the Americas: transnationalizing "fast" social policy', *Geoforum* 41: 195–208 <http://dx.doi.org/10.1016/j.geoforum.2010.01.001>.

Perreault, T. (2003) 'Making space: community organization, agrarian change, and the politics of scale in the Ecuadorian Amazon', *Latin American Perspectives* 30: 96–121 <www.jstor.org/stable/3184967>.

Reed, M.G. and Bruyneel, S. (2010) 'Rescaling environmental governance, rethinking the state: a three-dimensional review', *Progress in Human Geography* 34: 646–53 <http://dx.doi.org/10.1177/0309132509354836>.

Regmi, B.R., Morcrette, A., Paudyal, A., Bastakoti, R. and Pradhan, S. (2010) *Participatory Tools and Techniques for Assessing Climate Change Impacts and Exploring Adaptation Options: A Community Based Tool Kit for Practitioners,* Kathmandu: DfID Livelihoods and Forestry Programme.

Ribot, J. (2011) 'Vulnerability before adaptation: toward transformative climate action', *Global Environmental Change* 21: 1160–2 <http://dx.doi.org/10.1016/j.gloenvcha.2011.07.008>.

Sikor, T., Stahl, J., Enters, T. et al. (2010) 'REDD-plus, forest people's rights and nested climate governance', *Global Environmental Change* 20: 423–5 <http://dx.doi.org/10.1016/j.gloenvcha.2010.04.007>.

Smit, B. and Wandel, J. (2006) 'Adaptation, adaptive capacity and vulnerability', *Global Environmental Change* 16: 282–92 <http://dx.doi.org/10.1016/j.gloenvcha.2006.03.008>.

Smith, N. (2004) 'Scale bending and the fate of the national', in Sheppard, E. and McMaster, R. (eds), *Scale and Geographic Inquiry: Nature, Society, and Method,* Oxford: Wiley-Blackwell, pp. 193–212.

Staeheli, L.A. (2008) 'More on the "problems" of community', *Political Geography* 27: 35–9.

Triscritti, F. (2013) 'Mining, development and corporate–community conflicts in Peru', *Community Development Journal* 48: 437–50 <http://dx.doi.org/10.1093/cdj/bst024>.

Van den Berg, R.D. and Feinstein, O. (2009) *Evaluating Climate Change and Development,* Washington, DC: World Bank.

van Immerzeel, W.H.M. (2006) *Poverty, How to Accelerate Change: Experience, Results and Focus of an Innovative Methodology from Latin America,* Cusco: Dexcel.

Wiggins, S. (2009) *CEDRA: Climate Change and Environmental Degradation Risk and Adaptation Assessment,* Teddington, UK: Tearfund.

Yates, J.S. (2012) 'Uneven interventions and the scalar politics of governing livelihood adaptation in rural Nepal', *Global Environmental Change* 22: 537–46 <http://dx.doi.org/10.1016/j.gloenvcha.2012.01.007>.

Young, O.R. (2006) 'Vertical interplay among scale-dependent environmental and resource regimes', *Ecology and Society* 11: 27–44 <www.ecologyandsociety.org/vol11/iss1/art27/>.

About the author

Julian Yates is a doctoral candidate in the Department of Geography at the University of British Columbia in Vancouver, Canada. Julian has 10 years of experience working in the development field with NGOs and in academia, and his current research focuses on the intersection between institutional change, livelihood adaptation and hybrid forms of socioeconomic development in the Andes.

CHAPTER 3

A natural focus for community-based adaptation

Hannah Reid

Biodiversity – the variety of all life, from genes and species to ecosystems – is intimately linked to climate change. Climate change impacts will affect natural systems, many of which are also key carbon sinks. Biodiversity maintenance and sustainable natural resource management can also help people adapt to the adverse impacts of climate change, especially the world's poorest, who are also disproportionately reliant on natural resources for their subsistence, well-being and livelihoods. This chapter explains the concept of ecosystem-based approaches to adaptation (EbA) and describes the key differences and synergies between EbA and community-based adaptation (CBA). While both are relatively new concepts, they can learn much from older disciplines such as community-based natural resource management and disaster risk reduction. Key emerging issues are explained, including the need for stronger evidence on EbA, the need to integrate social and ecological disciplines in adaptation planning, and the importance of genuine community-led approaches.

Keywords: biodiversity, resilience, ecosystem-based approach, community-based natural resource management, disaster risk reduction, ecosystem services

Biodiversity – the variety of all life, from genes and species to ecosystems – is intimately linked to climate change. The effects that climate change will have on biodiversity are increasingly well known. Changes in species' abundance and distribution have already been observed, as have shifts in ecosystem boundaries and in reproductive cycles and growing seasons. Natural fire regimes are changing, as are water catchment processes. Changes to the complex ways in which species interact (predation, pollination, competition and disease) are altering the composition of plant and animal communities, and some species have already become extinct (Booth, 2012; McCarthy, 2012; Reid, 2006; Thomas et al., 2004). The Intergovernmental Panel on Climate Change expects further changes to biodiversity and natural systems as climate change proceeds (Fischlin et al., 2007), and the need to deal with this challenge is acknowledged at the highest levels. Indeed, the ultimate objective of the United Nations Framework Convention on Climate Change (UNFCCC) is, among other things, to stabilize atmospheric greenhouse gas concentrations 'within a time frame sufficient to allow ecosystems to adapt naturally to climate change' (UN, 1992).

http://dx.doi.org/10.3362/9781780447902.003

The importance of natural systems such as forests and peatlands in absorbing and storing carbon dioxide and other greenhouse gases, and hence mitigating climate change, is also well documented. The role that biodiversity and natural resources have in helping people adapt to the adverse impacts of climate change has received less attention.

The world's poorest people will be worst hit by climate change because they live in vulnerable areas and have the least capacity to cope. Poor people are also disproportionately reliant on natural resources such as timber, fish, grazing and wild medicines for their subsistence, well-being and livelihoods. As climate change impacts worsen and the need to help those who are most vulnerable grows, sustainable management of these natural resources, and maintenance of genetic, species and ecosystem diversity, will therefore play a key role in helping them adapt to future climate change impacts (Reid, 2011).

It is therefore perhaps no surprise that ecosystem-based approaches to adaptation (EbA) have received increasing interest in policy circles and among practitioners in recent years. There are important differences between community-based adaptation (CBA) and EbA, but both have the end goal of increasing the ability of vulnerable people to adapt to climate change. This chapter explains our current understanding of what EbA are and describes the key differences and synergies between EbA and CBA. It argues that CBA cannot afford to neglect EbA and that in many instances a strong understanding of ecosystems and natural systems can dramatically increase CBA effectiveness. Indeed, investing in sustainable biodiversity and ecosystem management could provide a way to help meet two of the biggest global challenges facing mankind – climate change and biodiversity loss. Biodiversity and ecosystem services are already the foundation of many successful adaptation strategies, especially for poor people, and many also deliver livelihood and climate change mitigation benefits.

Like CBA, EbA are conceptually relatively new and lack a long history of 'learning by doing' and scientific analysis behind them. Learning on EbA need not start from scratch, however; it can build on experience from older disciplines such as community-based natural resource management (CBNRM), drylands management, disaster risk reduction and agroecology. This chapter identifies information relevant to EbA that already exists in these fields of research and practice, using CBNRM as a particular example. Learning from CBNRM, it then describes some key lessons regarding the things EbA need to address if they are to gain widespread support.

Finally, the chapter identifies some of the key lessons and emerging issues for CBA and EbA that focus on the important role that ecosystems and ecosystem services play in adaptation. First is the fact that resilient diverse ecosystems can play a key role in supporting adaptation, and to this end the author proposes adopting an integrated approach to adaptation that combines the two disciplines of CBA and EbA and builds on their respective strengths. Scaling up is central if local adaptation initiatives are to comprise more than a handful of projects. The importance of taking an ecosystem approach to scaling up, one that complements efforts to scale up using social and political frameworks, is also emphasized, as well as the importance of maintaining a genuine community-led approach. Lastly, the chapter calls for more research to strengthen the scientific evidence base for EbA in order to better inform adaptation planning, and it describes the key knowledge gaps that need addressing in this field.

What is an ecosystem-based approach to adaptation?

EbA is defined by the United Nations Convention on Biological Diversity (CBD) as 'the use of biodiversity and ecosystem services to help people adapt to the adverse effects of climate change as part of an overall adaptation strategy' (CBD, 2009). This definition was later elaborated by the CBD to include 'sustainable management, conservation and restoration of ecosystems, as part of an overall adaptation strategy that takes into account the multiple social, economic and cultural co-benefits for local communities' (CBD, 2010). While some in the conservation sector see EbA as the adaptation of ecosystems in the face of climate change, the above definition makes it very clear that human adaptation is at the centre of EbA. Andrade et al. (2011) describe some key principles for EbA, many of which are common to CBA. They state that EbA:

- are about promoting the resilience of both ecosystems and societies, in other words promoting 'The ability of a social or ecological system to absorb disturbances while retaining the same basic structure and ways of functioning, the capacity for self-organization, and the capacity to adapt to stress and change' (Parry et al., 2007);
- promote multi-sectoral approaches;
- operate at multiple geographical and temporal scales;
- integrate flexible management structures that enable adaptive management;
- minimize trade-offs and maximize benefits with development and conservation goals to avoid unintended negative social and environmental impacts;
- are based on best available science and local knowledge, and foster knowledge generation and diffusion;
- are about resilient ecosystems, and using nature-based solutions at the service of people, especially the most vulnerable;
- are participatory, transparent, accountable and culturally appropriate, and actively embrace equity and gender issues.

Munroe et al. (2011) provide some examples of EbA: defence of coastal areas by maintaining or restoring coastal vegetation such as mangroves to reduce wave strength and therefore coastal flooding, coastal erosion and storm damage; sustainable management of wetlands and floodplains to maintain water flow and water quality in the face of changing rainfall regimes, and to provide floodwater reservoirs to reduce downstream flooding and important water stores in times of drought; conservation and restoration of forests and natural vegetation to stabilize slopes and regulate water flows, preventing flash flooding and landslides; and establishment of healthy and diverse agroforestry systems (the integration of food production into forests) to cope with changed climatic conditions by maintaining genetic diversity of crops and livestock and using indigenous knowledge of crop and livestock varieties.

How do ecosystem-based approaches to adaptation differ from community-based adaptation?

CBA has been defined as 'a community-led process, based on communities' priorities, needs, knowledge and capacities, which should empower people to plan for and cope with the impacts of climate change' (Reid et al., 2009). It adopts a people-centred rather than an environment-centred approach, which builds on various principles and processes linked to human rights-based approaches to development. These target the most vulnerable people and fully include them in all levels of adaptation planning and implementation. They understand and address local needs and ensure that adaptation activities do not inadvertently worsen vulnerability. Good CBA is as much about 'process' as it is about 'outcomes' (Reid and Schipper, 2014). Human rights-based approaches also seek to redress the power imbalances that make some people more vulnerable than others. Some CBA activities consider ecosystem goods and services when local people and livelihoods clearly depend on them, but this tends to be in the context of natural resources – such as forest products, water, agricultural yields or fish stocks – as opposed to complex ecosystems and the services they provide to reduce hazards and improve local resilience (Girot et al., 2012).

By contrast, EbA place an understanding of ecosystems at the centre of adaptation planning. They consider the natural resources described above, but also key ecosystem services such as pollination, regulation of the climate, genetic diversity and the connections and links between different natural systems. Just as CBA is informed by human rights-based approaches, EbA are informed by principles of the ecosystem approach to conservation that were adopted by the CBD in 2000 and endorsed by the World Summit on Sustainable Development in 2002. The ecosystem approach seeks to maintain ecosystem services by conserving ecosystem structure and functioning, recognizing that ecosystems have limits, undergo change and are interconnected.

Some of the perceived differences between EbA and CBA are due to the fact that each has evolved from a different set of practitioners. CBA has primarily been championed by the development community, and EbA by the environment and conservation community. These two groups have different values, institutional agendas and donor funding sources, but over the last two years they have found more in common with each other than differences between their ideologies would suggest. It is now increasingly common to find environment- and development-oriented organizations working together at project, programme and policy levels to support adaptation policy, planning and implementation.

Ecosystem-based and community-based approaches in practice: the many synergies

CBA and EbA each have their own specific emphasis, but, in practice, local adaptation activities tend to combine both approaches. For example, communities throughout the world have been using genetic diversity and traditional knowledge about native species to adapt to climate variability for generations, and increasingly this knowledge is proving valuable in the context of climate change adaptation (Swiderska et al., 2011). Likewise, mangroves are well-known coastal buffers, reducing the strength of waves before they reach the shore and thus protecting against cyclone damage in addition to sequestering carbon and providing a resource base for local livelihoods and income generation (see, for example, Mavrogenis and Kelman, 2012). Wetlands act as important floodwater reservoirs, and vegetation such as hedges protects agricultural land from excessive water or wind erosion in times of heavy rainfall or drought. Vegetation on hillsides reduces erosion and the risk of landslides when rain comes in heavy bursts.

Well-vegetated watersheds also slow the movement of rainfall to rivers, thus reducing flood risks downstream. Many of the urban poor of Jakarta, for example, live in low-lying slum areas along river banks and canals. In 2006, Mercy Corps began work to improve the lives and livelihoods of people living in one such area. In early 2007, however, Jakarta experienced its worst recorded flood, which inundated large parts of the city, sent over 422,000 fleeing their houses and led to many thousands of people falling ill with flood-related diseases. The flood was a result of heavy rain like that expected with climate change, exacerbated in part by deforestation on higher land south of the city that caused rapid water run-off. Mercy Corps – a humanitarian and development organization – realized that building community resilience to future climate change required action beyond the local slum areas where they had a physical presence. They now collaborate with conservation groups working in the mountain areas around Jakarta to conserve the forests and protect green spaces, both of which help reduce the rapid water run-off (Jeans et al., 2014).

Why community-based adaptation must not neglect ecosystem-based approaches to adaptation

The potential benefits of CBA may be undermined if broader ecosystem processes and services are not considered in planning and implementation. For example, building a dam may provide one community with water but adversely affect other communities downstream. Likewise, digging multiple wells may lower the overall water table in the region, adversely affecting neighbouring communities. Planning at the river basin or watershed level in order to secure adequate community water supplies can help avoid such maladaptation. Local responses need contextualizing in the wider landscape or ecosystem if they are to be effective (McCarthy, 2012). This is not always easy as the boundaries of ecosystems may not correspond to political or administrative boundaries.

Natural timescales also need considering in adaptation planning and, as with the natural spatial scales described above, these may not always correspond to the social or political timescales more usually adopted when planning a CBA project. For example, many coastal adaptation projects involve planting mangroves. In the longer term, however, mangroves will be affected by sea-level rise, and if there is no space for them to migrate inland to avoid being submerged by rising sea levels, then investments in short-term adaptation efforts will be wasted. Likewise, investments in new crops better adapted to changing climatic conditions could be wasted if changes in temperatures mean that the crops' main pollinators will not have hatched or will have completed the key (pollinating) part of their lifecycle by the time their pollination services are required.

CBA projects with strong natural resource components need to remember that ecosystems are not static. Many may not be able to provide the natural resources and services that people rely on and anticipate benefiting from in the future (Watson et al., 2012). In some instances, ecosystems might be damaged irreparably by climate change and other stressors if they are pushed beyond the limits at which they can function. For example, the frequency and severity of El Niño events are likely to increase due to climate change, and this could push many coral reefs beyond the threshold of recovery, with significant resultant economic losses from damage to fishing, tourism and livelihoods.

There is some evidence to suggest that EbA can be cost-effective and considerable evidence to suggest that they can generate a multitude of social, economic and environmental co-benefits (Doswald et al., 2014; Munroe et al., 2011, 2012). These include disaster risk reduction, livelihood sustenance and food security, carbon sequestration, sustainable water management and even a reduction in conflict over scarce resources. Such co-benefits are not always easy to account for in narrow analyses focusing on the immediate financial costs and benefits of a particular adaptation activity, but they are nonetheless important. In 1994, for example, the Vietnam Red Cross started to work with local communities to plant and protect mangrove forests in northern Vietnam. Nearly 12,000 hectares of mangroves were planted (and protected) at a cost of approximately $1.1 million, and the project ultimately saved $7.3 million a year in dyke maintenance. Storms and tropical cyclones are likely to increase in frequency and severity in the region due to climate change, and the adaptation benefits of these mangroves were apparent in 2000 when Typhoon Wukong hit the area. Project areas remained unharmed while neighbouring provinces suffered huge losses in lives, property and livelihoods. The Vietnam Red Cross estimated that some 7,750 families benefited from mangrove rehabilitation. Family members could also earn additional income from selling the crabs, shrimp, molluscs and seaweed that thrive in the mangroves and increase the protein in their diets. The mangroves also sequester carbon (IFRC, 2001).

The importance of incorporating ecosystems and ecosystem services into adaptation planning and policy making is gaining widespread recognition, and Jeans et al. (2014) provide some guidance on how to do this at a practical level (see Box 3.1). The Cancun Adaptation Framework, adopted under the UNFCCC in December 2010, recognizes the role of natural resource management as an adaptation action that increases resilience of socioeconomic and ecological systems (Decision 1/CP16. 14d). A number of national climate change policies and strategies and sector-based policies (for example on water, forests and coastal zone management) also recognize the role ecosystems play in adaptation. A study of the national adaptation programmes of action (NAPAs) of the least developed countries (LDCs), for example, shows that many LDCs recognize and prioritize the role that biodiversity, ecosystems and natural habitats play in adaptation. The study found that some 56 per cent of priority NAPA projects reviewed had significant natural resource components, and in Cape Verde, Eritrea, Sudan, Solomon Islands and Vanuatu, every NAPA project reviewed had a strong natural resource component (Reid et al., 2009). EbA are also now widely endorsed by a number of influential multinational and environmental organizations, including the International Union for the Conservation of Nature, the World Bank and the United Nations Environmental Programme.

Box 3.1 How to integrate ecosystems successfully into adaptation strategies and programmes, including community-based adaptation

- As well as working at the local community level, think and plan at the spatial scales at which ecosystems operate (for example, the watershed management scale) and engage with institutions operating at these other scales to create an enabling environment that can address ecosystem functions.
- Integrate endogenous (internal) and exogenous (external) knowledge so that local and traditional knowledge on natural resource or ecosystem management for livelihoods and human well-being is supported by appropriate exogenous technology.
- Maintain ecosystem services at multiple scales (for example, replanting deforested slopes, managing water resources on a catchment-wide scale with multi-stakeholder participation, and restoring and replanting mangroves).
- Increase ecosystem resilience (for example, by removing non-climate stressors and promoting ecosystem management and restoration).
- Facilitate ecosystem adaptation (recognizing that ecosystems and natural resources will not remain stationary in the face of climate change).
- Promote 'no regret' or 'low regret' locally appropriate nature-based solutions.
- Promote partnerships that bring together local communities and decision makers with development, conservation and disaster risk reduction communities of practice.
- Ensure that the full extent of ecosystem services and values is included in any assessment of costs and benefits for adaptation planning.
- Monitor and manage ecosystem change to support ongoing human and ecosystem resilience and adaptation.

Source: Jeans et al., 2014

Strengthening the scientific evidence base relating to ecosystem-based approaches to adaptation

The importance of considering ecosystems and the services they provide for any CBA activities is clear, and the growing body of institutional and policy support for the principles of EbA is encouraging. As with CBA, however, EbA are on relatively new ground in terms of learning and practice, and remain largely untested in the field. Adaptation is a long-term challenge but most planned projects using EbA are newly established and so cannot yet provide evidence regarding the long-term effectiveness of EbA as opposed to alternative adaptation responses.

There have been few scientific studies on the effectiveness of EbA, and only a limited number of reviews of the existing case studies. In particular there are very few studies providing an analysis of two comparable sites – one with and one without EbA – or a 'before and after' situation in the event of a dramatic climate change impact such as a cyclone. Similarly, there are few case studies that closely examine who benefits from EbA among vulnerable communities and groups and across broader scales. And when it comes to evaluating value for money, few case studies using EbA provide a quantified economic assessment. Such economic data can be difficult to obtain but are likely to provide the biggest justification to decision makers when it comes to adopting a new approach.

Existing evidence regarding EbA largely comes from anecdotes which, although informative, provide rather limited insight in terms of measuring and evaluating the effectiveness of EbA, especially when compared with technical or structural adaptation measures (Munroe et al., 2011; Reid, 2011). Indeed, much 'grey literature' focuses on trying to advocate for the adoption of EbA without including obvious measures of effectiveness (Munroe et al., 2012).

This shortage of scientific evidence for EbA does not mean that all practice and learning on EbA should grind to a halt. It is clear, however, that existing action learning should proceed in conjunction with greater academic rigour in order to better define the conditions under which EbA are effective, their limits and thresholds, their costs and benefits, and to develop improved guidance for their application.

Although the use of EbA is a relatively new field, much can be learned from related scientific fields of study, such as disaster risk reduction and dryland management, CBNRM and agroecological approaches to resilience (IAASTD, 2009). Many of these have many decades of relevant documented experience on sound environmental management that long precedes any political interest in climate change (Chishakwe et al., 2012; Munroe et al., 2011; Reid, 2011). Likewise, making use of ecosystems and their services to adapt to current climate variability and hazards (rather than climate change) – for example, restoring rivers to alleviate flooding – is common practice.

Learning from community-based natural resource management

CBNRM has evolved over 30 years and now has an established set of principles and approaches that could contribute much to EbA (Chishakwe et al., 2012). Learning from CBNRM can also tell us more about the processes by which new approaches to an environmental or development problem become mainstreamed (or not) into the daily discourses of politicians and practitioners, and about how such approaches gain popularity or fall from grace. This could smooth the journey ahead for the newer, related disciplines of EbA and CBA.

In its heyday, CBNRM was heralded as an alternative approach to the kinds of conservation activities favoured at the time involving national parks, armed rangers, fences and the separation of natural resources from the local people who had hitherto relied on them (IIED, 1994). Many CBNRM initiatives, however, were in fact externally initiated or imposed, and genuine systematic attempts by conservation authorities to adopt participatory planning methods were few and far between (Adams and Hulme, 2001; Murphree, 2000; Pimbert and Pretty, 1995; Reid et al., 2004).

The first key lesson from this is that we must be careful to clearly define exactly what we mean by these EbA and CBA, and assess whether the principles developed so far are actually being applied (Chishakwe et al., 2012). At the six international CBA conferences held to date, it has been apparent that many organizations are opportunistically seeking to label their development work as CBA in order to capitalize on potential advantages (for example, climate change donor funding). These activities, however, do not take a climate vulnerability analysis as their starting point. Similarly, some interventions that claim to be EbA are actually traditional conservation projects or CBNRM projects that merely deal with conservation or resource challenges as opposed to specific climate change-related vulnerabilities. While it is likely that many of these activities have much value in terms of learning about increasing resilience to climate change impacts and informing adaptation approaches, repackaging 'old wine in new bottles' in this way is unwise. Experiences with CBNRM warn us that the concept of EbA (and indeed CBA) may be discredited by activities labelled as such that 'fail', but that were actually not EbA (or CBA) in the first place.

Secondly, the practice of EbA needs more analytical rigour in order to assess whether it really works or not. Despite the urgency in terms of action that climate change scenarios tell us is needed to tackle this global threat, EbA must not try to run before they can walk. It will take time to gather the robust evidence needed to objectively explain the merits and also the limitations of EbA. As with CBA, the concept may have energy and momentum associated with it now, but in view of the changing winds of donor funding and trends in development assistance – and climate change is a very political, fast-moving arena at present – a stronger evidence base is needed if EbA are to stand the test of time. Many of those currently involved in activities using EbA are practitioners working in or close to the field, but the skills of these people need to be complemented by an objective analysis of EbA strengths and limitations. Without this analysis the concept could lose credibility before it has had the chance to genuinely prove itself or otherwise.

Thirdly, over 30 years' experience with CBNRM have taught us the importance of addressing the institutional and policy environment in which EbA operate if they are to successfully take root. CBNRM was initially seen primarily as a conservation approach, which later emerged to have rural development co-benefits, but 'It is now viewed as an institutional or organisational development programme whereby natural resources are utilised to economically empower local people' (Chishakwe et al., 2012). We can perhaps expect a similar trajectory for EbA (and CBA). Chishakwe et al. (2012) expand on this by explaining how CBNRM's most remarkable attribute has arguably been the processes and institutions it has established in order to achieve its results. This includes the creation of space for the direct and practical involvement of communities; the devolution of power from central government to communities recognized by policy and law; the establishment of mechanisms to ensure the provision of tangible benefits for communities from conservation initiatives; and the opportunity this provides for replication and diversification to other sectors. Central to this is engagement with effective, legitimate local institutions that incorporate – or are based on – appropriate traditional forms of governance in order to secure local legitimacy. Chishakwe et al. (2012) argue that CBA must ensure inclusive institutional arrangements are in place, and 'create space for all relevant stakeholders – such as elected representatives, community members, non-governmental organizations (NGOs) and the private sector – to participate'. The same is true for EbA, whose success or otherwise is likely to hinge on the institutional and policy environment in which EbA operate, because this will determine whether local communities are in control of their adaptation activities or not.

The last lesson from CBNRM discussed here is the importance of community 'incentives'. Chishakwe et al. (2012) describe how 'in order to successfully implement a community-led project whose benefits are only realized in the long term, it is important to have interim (incentive) mechanisms that compensate for any short-term costs'. They describe the importance of cash and non-cash benefits for CBNRM and argue that adaptation practitioners and theorists need to consider such incentives if communities are to be sufficiently motivated to adopt adaptation actions. Without mechanisms to compensate the community for short-term losses (in terms of the time and resources spent on adaptation activities that could be spent on other livelihood activities), it will be difficult to achieve the intended adaptation project objectives.

For many communities, the benefits from investing in a comparatively abstract issue such as 'adaptation', where there is often considerable uncertainty regarding the risks ahead, are less tangible than those from addressing more immediate and certain health, nutritional or livelihood concerns. Poor people often value the present more than the future, so activities must have direct and immediate benefits for them. This is not necessarily compatible with the long time frames needed to allow effective adaptation to occur. CBNRM initiatives provided short-term cash in hand to local communities from wildlife concessions, but in the case of local-level adaptation activities it is less clear where funding might come from. There are funds emerging from the global climate change policy arena, for example the Adaptation Fund has demonstrated some early successes in terms of providing financial support for adaptation activities that benefit local-level adaptation activities, but international and national systems for financing local adaptation activities are mostly absent or at best in their infancy.

Key emerging issues for community-based adaptation

This chapter now proposes four key lessons and emerging issues for CBA (and EbA) practitioners to take forward.

Resilient, diverse ecosystems will support human adaptation

The first clear lesson for CBA is that well-managed, stable, diverse ecosystems can make a significant contribution to local adaptation efforts. For example, maintaining a wide diversity of species and genetic diversity can facilitate the emergence of species and genotypes that are better adapted to shifts in climatic conditions and could well have important adaptation benefits in the future. *The Economics of Ecosystems and Biodiversity* report (European Communities, 2008) explains this further:

> The security value of biodiversity can be compared with financial markets. A diverse portfolio of species stocks, as with business stocks, can provide a buffer against fluctuations in the environment (or market) that cause declines in individual stocks. This stabilizing effect of a 'biodiverse' portfolio is likely to be especially important as environmental change accelerates with global warming and other human impacts.

Ecosystem resilience to climate change is generally higher if the system is in good condition and non-climate stressors such as habitat destruction, overharvesting of resources and pollution are minimized (Hansen et al., 2003). Promoting healthy ecosystems and reducing non-climate stressors on these ecosystems will help maintain ecosystem services and thus support human adaptation (Chishakwe et al., 2012). For example, reforestation and conservation of intact forests, maintaining or restoring connectivity between natural spaces, avoiding overuse of resources and reducing the risk of forest fires can help increase forest resilience to climate change. This in turn helps to ensure continued access to natural forest resources that support people's livelihoods and reduce their vulnerability to shocks and other climate change impacts. It can also reduce the risk of disaster events. For example, forest cover can stabilize hillsides, thus reducing the chances of landslides that may be triggered by intense rainstorms.

It is not easy to predict what climate change impacts to expect in any single location over the coming years. It is therefore advisable for adaptation strategies to adopt a precautionary approach and strive to promote healthy ecosystems and reduce non-climate stressors in order to ensure that they can support adaptation activities where necessary in the future (Jeans et al., 2014).

Adopting an integrated approach to adaptation

Reflecting on the commonalities between CBA and EbA, several leading conservation and development practitioners have been arguing for the promotion of an 'integrated approach' to adaptation that breaks down the artificial divide between CBA and EbA and builds on the strengths of both concepts (Girot et al., 2012). They define this integrated approach as 'adaptation planning and action that adheres both to human rights-based principles and principles of sound environmental management, recognising their interdependent roles in successfully managing climate variability and long-term change'. They stress the interconnectedness of ecosystems, biodiversity, poverty and adaptation to climate change and argue that such integrated approaches have a better chance of forcefully addressing the shortcomings of the mainstream top-down, hard infrastructure-based approaches to adaptation espoused by many major international financial institutions than either EbA or CBA would have alone (Girot et al., 2012). Experience on the ground in Tonga also seems to reinforce the fact that a 'portfolio' approach linking CBA, EbA, livelihoods and in some instances even an infrastructure-based approach to adaptation will likely provide the best outcomes (Mavrogenis and Kelman, 2012).

The adoption of integrated approaches in practice will require learning and the application of new methodologies by both EbA and CBA practitioners. CBA practitioners will need a better understanding of biodiversity, ecosystems and ecosystems services. They need to move beyond the perception of ecosystem goods and services as a set of static, finite natural resources, towards a fuller understanding of ecological complexity and interdependence. They need to learn that ecosystems are likely to change over time due to climate change and other stressors and that in some instances action at the appropriate ecosystem scale in addition to (or in some cases instead of) at the appropriate geographical, political or social scale may enhance overall adaptation benefits. Planning needs to incorporate the latest science on climate change impacts on environmental as well as social systems; vulnerability analyses should have environmental as well as social components; and monitoring and evaluation systems for adaptation need to include issues relating to environmental integrity (Girot et al., 2012). As a bare minimum, adaptation interventions should not undermine the health or resilience of ecosystems, and should 'aim to minimise harm to ecosystems and biodiversity. This should help avoid maladaptation and negative repercussions on people's ability to adapt and help ensure resilience into the future' (Reid et al., 2009).

Improvements in learning are not needed by CBA practitioners alone. Those using EbA often have a background in conservation and lack the skills and knowledge needed to ensure that adaptation planning is truly bottom-up and participatory in nature. They need to get better at working with communities to identify local priorities, capture and build on traditional knowledge, and strengthen local capacity to act. Just as CBA practitioners need to get better at addressing ecological complexity in their activities, EbA practitioners need to go beyond seeing 'the community' as a homogeneous group of individuals in one location and better address the socioeconomic complexities and differential vulnerabilities that characterize communities everywhere.

Better integration at the practitioner level needs complementing with improved integration at the institutional and policy level. To date, CBA and EbA have primarily been championed by development and conservation practitioners respectively. These two groups have different values and funding sources and also different institutional and policy entry points for securing change. To the extent that national governments have engaged with either approach, it is usual for each to be housed under different ministries and national policy processes. For example, climate change (and hence CBA) often falls under the remit of government meteorological departments, whereas conservation and environment management (and hence EbA) are more often housed under the activities of environmental ministries. Improved cross-sectoral learning and partnerships between these different institutions and policy processes are needed if integrated approaches are to be promoted successfully.

Scaling up

National and local decision makers have available to them a range of adaptation pathways, one of which is to support large-scale infrastructure projects, such as dams and dykes, with the objective of reducing people's vulnerability to climate change hazards. Such projects are often favoured because they provide measurable outcomes within an electoral cycle, impressive political kudos and great photo opportunities, and they fit within prevailing discourses on technology and development. However, evidence from decades of experience in development and disaster risk reduction shows that these large-scale, hard infrastructure interventions are very costly and often neglect the needs of the most vulnerable, in some cases making life much worse for local communities. They can also lead to maladaptation: for example, flood barriers can constrain ecologically important processes such as annual river flooding and coastal sediment transport; and, while a new dam might assist with water shortages in one area, it might also change the natural flooding regime downstream and thus damage agricultural production and local livelihoods (Girot et al., 2012; Munroe et al., 2011).

Doswald et al. (2014) observe that EbA deserve greater policy attention and political support to ensure that they reach their full potential and help provide alternative adaptation options to these large-scale infrastructure projects. Compared with these engineered 'solutions' to climate change, relatively little attention has been paid to non-structural nature-based alternatives or to community-based strategies for managing natural resources, despite the fact that they may be able to provide cheaper, more sustainable solutions with multiple co-benefits (Jeans et al., 2014).

To this end, scaling up existing activities is a key challenge for EbA, and indeed for CBA. Most initiatives to date are small, stand-alone activities that operate at a localized scale with support from conservation or development NGOs. Such activities are proliferating rapidly, but if they are to benefit more than just those involved in these specific isolated initiatives, planning needs to extend beyond the project level to the programme and policy level (Schipper et al., 2014). As Steele et al. (2008) explain in the context of development interventions more generally: 'The challenge is to scale up and ultimately support thousands and even millions of poor people to raise themselves out of poverty.'

More often than not, such scaling up requires systematic mainstreaming into government structures, policies, laws and planning processes. In Nepal, for example, local adaptation plans of action have been embedded within the National Adaptation Programme of Action to ensure that bottom-up adaptation planning is mainstreamed into government policy and planning processes. Embedding EbA in local, regional and national government policies, processes and institutions in this way will support the wide-scale replication of these approaches and can also provide a way to tackle the issue of incentives, because such structures can channel emerging financial support for EbA activities (from mechanisms in the international climate change policy arena, donors or, increasingly, national budgets). Addressing the issue of broad-scale incentives is important, because without doing so EbA will be limited to small-scale 'pilot' projects with little chance of benefiting more than a handful of the millions who need help adapting to climate change (Hartmann and Linn, 2008).

Integrating EbA into adaptation policy processes such as national adaptation programmes of action, national adaptation plans or city development plans is one way to scale up the adoption of EbA (Munroe et al., 2011), but in most cases these are quite top-down government-led approaches to adaptation planning. Experiences from CBNRM remind us of the importance of establishing a policy and institutional environment that also supports the direct involvement of the community affected. Ensuring that local community involvement is legitimate might require recognition of existing traditional institutional structures. Traditional leadership plays an important role in symbolizing community ownership over local projects, and relationships of trust are an important factor in promoting communal proprietorship over initiatives and hence ensuring their success (Chishakwe et al., 2012).

Just as local approaches to adaptation need to be integrated into broader social and political policy and planning frameworks in order to be scaled up effectively, they also need to be integrated into larger-scale issues relating to ecosystem functioning, such as watersheds, natural coastal defence systems and sustainable forest management plans. Without this, scaling up could end up being maladaptive in many natural and urban environments. Ensuring that there is harmonization between scales of critical ecosystem function and also between political and social scales of intervention can be challenging, because it is common for each of these to have quite different boundaries. Scaling up across social, political and ecological scales and across disciplines will require collaborative partnerships that facilitate social learning between a wide variety of different stakeholders who work in different sectors and at different scales and may not be used to cross-disciplinary and collaborative ways of working (Jeans et al., 2014). For example, effective adaptation planning and management at the watershed or river basin scale may require national government bodies in several different countries to work both together and with multiple communities (Girot et al., 2012).

Improving the scientific evidence base

While anecdotes about EbA abound, the scientific evidence base for EbA is not yet strong, and policy makers need more than anecdotes on which to base their decisions. A recent review of existing evidence for EbA in the scientific and grey literature identified some specific gaps that particularly need addressing (Munroe et al., 2011; Doswald et al., 2014). A greater understanding of the thresholds, boundaries and tipping points across a range of EbA in varying climatic zones is needed as these determine whether ecosystems can continue to fulfil their required functions under particular conditions, and what type of EbA might best be applied in different situations. Ecosystems have limits beyond which they cannot function effectively, and these limits are complex and not always predictable (CBD, 2009). In many cases it is also unclear exactly how climate change will affect specific ecosystems, and if and when it will tip them beyond their coping limits. In some instances, ecosystem responses to climate change appear to be non-linear. Greater consideration of the temporal and spatial aspects of EbA effectiveness is also needed, as these aspects will inform whether EbA may or may not be successful under certain conditions.

More detailed comparisons between EbA and alternative adaptation strategies are needed, taking into account the social, environmental and economic costs as well as benefits. Current literature tends to advocate for EbA and report on their benefits, but learning from mistakes and negative outcomes can often be more informative for future planning. More studies comparing EbA with other adaptation options might help address the perennial problem of non-reporting on negative results and provide useful information to policy makers who require information on the social, environmental and, most importantly, economic costs and benefits of alternative options to facilitate effective decision making (Munroe et al., 2011; Doswald et al., 2014).

The published evidence base is skewed towards particular climate change impacts and particular types of ecosystems and ecosystem services. For example, literature on how EbA have been used to address sea-level rise and associated impacts, as well as to control erosion and reduce risk from natural disasters, is more extensive. In terms of sectors, the recent review of existing evidence found that most of the interventions using EbA addressed agriculture and the loss of biomass cover and productivity as their main target, with water, coastal protection, the provision of alternative livelihoods, forestry and urban areas also important targets. Evidence regarding the effectiveness of EbA in response to other climate impacts and in other sectors is weaker (Doswald et al., 2014).

A further gap in the evidence base for EbA is on social assessments (Munroe et al., 2011; Doswald et al., 2014). These are crucial 'because the impacts of climate change often disproportionately affect the most vulnerable communities and groups' (UNFCCC, 2010) and yet the literature on EbA provides little evidence about which particular communities and vulnerable groups are likely to benefit or suffer from the application of these approaches.

Learning from CBNRM stresses the importance of being very clear about what EbA are in order not to discredit these approaches before they have been properly tried and tested. It is therefore a concern that the review of the existing evidence for EbA found a lack of consistent use of terminology in published literature to date (Doswald et al., 2014). Without consensus on what successful or effective adaptation actually means, measuring it and monitoring it are also challenging (Spearman and McGray, 2011). A number of agencies are starting to address the issue of monitoring and evaluating climate change adaptation more generally (Brooks et al., 2011). Such monitoring and evaluation are in their infancy with EbA, but this will need addressing in due course (Doswald et al., 2014).

Efforts should be made to ensure that learning from past and current implementation efforts relating to EbA are captured and used to inform future activities. Learning must continue hand in hand with 'doing' for the potential benefits of EbA to be realized.

Conclusions

This chapter has stressed the importance of biodiversity, ecosystems and ecosystem services in the context of helping people adapt to the adverse impacts of climate change, especially the world's poorest, who are often most vulnerable to climate change impacts and most reliant on natural resources. It has argued that CBA cannot afford to neglect EbA in its planning and implementation, and that integrating both disciplines and building on their respective strengths could potentially maximize adaptation benefits, avoid maladaptation and provide a multitude of other livelihood, social, economic and environmental co-benefits.

More research to strengthen the scientific evidence base for EbA is needed, however, and the chapter describes some of the key knowledge gaps identified to date. Despite this, much can be learned from related disciplines, and the chapter describes four key lessons from CBNRM – with its longer history of theory and practice – that those engaged with EbA should consider. Firstly, the EbA concept needs to be defined clearly and an assessment made regarding whether the principles developed so far are being applied genuinely or not. Secondly, the practice of EbA needs more analytical rigour in order to assess whether it really works or not. Thirdly, the institutional and policy environment in which EbA operate needs prioritizing. Lastly, community 'incentives' need to be incorporated into all activities using EbA.

References

Adams, W. and Hulme, D. (2001) 'Conservation and communities: changing narratives, policies and practices in African conservation', in Hulme, D. and Murphree, M. (eds), *African Wildlife and Livelihoods: The Promise and Performance of Community Conservation*, Oxford: James Currey.

Andrade, A., Córdoba, R., Dave, R., Girot, P., Herrera-F., B., Munroe, R., Oglethorpe, J., Paaby, P., Pramova, E., Watson, E. and Vergar, W. (2011) *Draft Principles and Guidelines for Integrating Ecosystem-based Approaches to Adaptation in Project and Policy Design: A Discussion Document*, Turrialba, Costa Rica: Centro Agronómico Tropical de Investigación y Enseñanza (CATIE).

Booth, T.H. (2012) 'Biodiversity and climate change adaptation in Australia: strategy and research developments', *Advances in Climate Change Research* 3(1): 12–21.

Brooks, N., Anderson, S., Ayers, J., Burton, I. and Tellam, I. (2011) *Tracking Adaptation and Measuring Development*, IIED Climate Change Working Paper No. 1, London: International Institute for Environment and Development (IIED).

CBD (2009) *Connecting Biodiversity and Climate Change Mitigation and Adaptation: Report of the Second Ad Hoc Technical Expert Group on Biodiversity and Climate Change*, CBD Technical Series No. 41, Montreal, Canada: Secretariat of the Convention on Biological Diversity (CBD).

CBD (2010) *Decision Adopted by the Conference of the Parties to the Convention on Biological Diversity at its Tenth Meeting: X/33 Biodiversity and Climate Change*, UNEP/CBD/COP/DEC/X/33, Nagoya: Convention on Biological Diversity (CBD) <www.cbd.int/doc/decisions/cop-10/cop-10-dec-33-en.pdf> [accessed 3 October 2013].

Chishakwe, N., Murray, L. and Chambwera, M. (2012) *Building Climate Change Adaptation on Community Experiences: Lessons from Community-based Natural Resource Management in Southern Africa*, London: International Institute for Environment and Development.

Doswald, N., Munroe, R., Reid, H., et al. (2014) 'Effectiveness of ecosystem-based approaches for adaptation: review of the evidence-base', *Climate and Development* (forthcoming).

European Communities (2008) *The Economics of Ecosystems and Biodiversity: An interim report*, Bonn: TEEB.

Fischlin, A., Midgley, G.F., Price, J.T., Leemans, R., Gopal, B., Turley, C., Rounsevell, M.D.A., Dube, O.P., Tarazona, J. and Velichko, A.A. (2007) 'Ecosystems, their properties, goods, and services', in Parry, M.L., Canziani, O.F., Palutikof, J.P., van der Linden, P.J. and Hanson, C.E. (eds), *Climate Change 2007: Impacts, Adaptation and Vulnerability. Contribution of Working Group II to the Fourth Assessment Report of the Intergovernmental Panel on Climate Change*, Cambridge: Cambridge University Press.

Girot, P., Ehrhart, C., Oglethorpe, J., Reid, H., Rossing, T., Gambarelli, G., Jeans, H., Barrow, E., Martin, S., Ikkala, N. and Phillips, J. (2012) *Integrating Community and Ecosystem-based Approaches in Climate Change Adaptation Responses*, ELAN, unpublished <http://elanadapt.net/sites/default/files/siteimages/elan_integratedapproach_15_04_12_0.pdf > [accessed 3 October 2013].

Hansen, L.J., Biringer, J.L. and Hoffman, J.R. (2003) *Buying Time: A User's Manual to Building Resistance and Resilience to Climate Change in Natural Systems*, Berlin: World Wildlife Fund.

Hartmann, A. and Linn, J.F. (2008) *Scaling Up: A Framework and Lessons for Development Effectiveness from Literature and Practice*, Working Paper 5, Washington, DC: Wolfenson Center for Development, The Brookings Institute.

IAASTD (2009) *Agriculture at a Crossroads: International Assessment of Agricultural Knowledge, Science and Technology for Development*, Washington, DC: Island Press.

IFRC (2001) *Coastal Environmental Protection: A Case Study of the Vietnam Red Cross*, Geneva: International Federation of Red Cross and Red Crescent Societies (IFRC).

IIED (1994) *Whose Eden? An Overview of Community Approaches to Wildlife Management. A Report to the Overseas Development Administration of the British Government*, London: International Institute for Environment and Development (IIED).

Jeans, H., Oglethorpe, J., Phillips, J. and Reid, H. (2014) 'The role of ecosystems in climate change adaptation: lessons for scaling up', in Schipper, E.L.F., Ayers, J., Reid, H., Huq, S. and Rahman, A. (eds), *Community Based Adaptation to Climate Change: Scaling it up*, London: Routledge.

Mavrogenis, S. and Kelman, I. (2012) 'Lessons from local initiatives on ecosystem-based climate change work in Tonga', in Renaud, F., Estella, M. and Sudmeier, K. (eds), *The Role of Ecosystems in Disaster Risk Reduction: From Science to Practice*, Tokyo: United Nations University Press.

McCarthy, P.D. (2012) 'Climate change adaptation for people and nature: a case study from the U.S. southwest', *Advances in Climate Change Research* 3(1): 22–37.

Munroe, R.N., Doswald, N., Roe, D., Reid, H., Giuliani, A., Castelli, I. and Möller, I. (2011) *Does EbA Work? A Review of the Evidence on the Effectiveness of Ecosystem-based Approaches to Adaptation*, Research Highlights, Cambridge: BirdLife International, United Nations Environment Programme World Conservation Monitoring Centre (UNEP-WCMC), International Institute for Environment and Development.

Munroe, R., Roe, D., Doswald, N., Spencer, T., Möller, I., Vira, B., Reid, H., Kontoleon, A., Giuliani, A., Castelli, I. and Stephens, J. (2012) 'Review of the evidence base for ecosystem-based approaches for adaptation to climate change', *Environmental Evidence* 1: 13 <www.environmentalevidencejournal.org/content/1/1/13>.

Murphree, M.W. (2000) *Community-Based Conservation: Old Ways, New Myths and Enduring Challenges*. Key Address at the Conference on African Wildlife Management in the New Millennium. College of African Wildlife Management, Mweka, Tanzania, 13–15 December 2000.

Parry, M.L., Canziani, O.F., Palutikof, J.P., van der Linden, P.J. and Hanson, C.E. (eds) (2007) *Climate Change 2007: Impacts, Adaptation and Vulnerability. Contribution of Working Group II to the Fourth Assessment Report of the Intergovernmental Panel on Climate Change*, Cambridge: Cambridge University Press.

Pimbert, M.P. and Pretty, J.N. (1995) *Parks, People and Professionals: Putting 'Participation' into Protected Area Management*, Discussion Paper 57, Geneva: United Nations Research Institute for Social Development.

Reid, H. (2006) 'Climatic change and biodiversity in Europe', *Conservation and Society* 4(1): 84–101.

Reid, H. (2011) *Improving the Evidence for Ecosystem-based Adaptation*, IIED Opinion: Lessons from Adaptation in Practice series, London: International Institute for Environment and Development (IIED).

Reid, H., Alam, M., Berger, R., Cannon, T., Huq, S. and Milligan, A. (2009) 'Community-based adaptation to climate change: an overview', *Participatory Learning and Action* 60: 11–33.

Reid, H. and Schipper, E.L.F. (2014) 'Upscaling community-based adaptation: an introduction to the edited volume', in Schipper, E.L.F., Ayers, J., Reid, H., Huq, S. and Rahman, A. (eds), *Community Based Adaptation to Climate Change: Scaling it up*, London: Routledge.

Reid, H., Fig, D., Magome, H. and Leader-Williams, N. (2004) 'Co-management of contractual national parks in South Africa: lessons from Australia', *Conservation and Society* 2(2): 377–409.

Reid, H., Phillips, J. and Heath, M. (2009) *Natural Resilience: Healthy Ecosystems as Climate Shock Insurance*, IIED Briefing, December, London: International Institute for Environment and Development (IIED).

Schipper, E.L.F., Ayers, J., Reid, H., Huq, S. and Rahman, A. (2014) *Community Based Adaptation to Climate Change: Scaling it up*, London: Routledge.

Spearman, M. and McGray, H. (2011) *Making Adaptation Count: Concepts and Options for Monitoring and Evaluation of Climate Change Adaptation*, Bonn: German Agency for International Cooperation (GIZ).

Steele, P., Fernando, N. and Weddikkara, M. (2008) *Poverty Reduction that Works: Experience of Scaling Up Development Success*. London and Colombo: Earthscan, with United Nations Development Programme (UNDP) Regional Centre.

Swiderska, K., Song, Y., Li, J., Reid, H. and Mutta, D. (2011) *Adapting Agriculture with Traditional Knowledge*, IIED Briefing Paper, October, London: International Institute for Environment and Development (IIED).

Thomas, C.D., Cameron, A., Green, R.E., Bakkenes, M., Beaumont, L.J., Collingham, Y.C., Erasmus, B.F.N., Ferreira de Siqueira, M., Grainger, A., Hannah, L., Hughes, L., Huntley, B., van Jaarsveld, A.S., Midgley, G.F., Miles, L., Ortega-Huerta, M.A., Peterson, A.T., Phillips, O.L. and Williams, S.E. (2004) 'Extinction risk from climate change', *Nature* 427: 145–8

UN (1992) *United Nations Framework Convention on Climate Change*, New York: United Nations (UN).

UNFCCC (2010) *Synthesis Report on Efforts Undertaken to Assess the Costs and Benefits of Adaptation Options, and Views on Lessons Learned, Good Practices, Gaps and Needs*, FCCC/SBSTA/2010/3, Bonn: United Nations Framework Convention on Climate Change (UNFCCC) <http://unfccc.int/resource/docs/2010/sbsta/eng/03.pdf> [accessed 3 October 2013].

Watson, J.E.M., Rao, M., Ai-Li, K. and Yan, X. (2012) 'Climate change adaptation planning for biodiversity conservation: a review', *Advances in Climate Change Research* 3(1): 1–11.

About the author

Dr Hannah Reid is a consulting researcher, currently working with the Climate Change Group at the International Institute for Environment and Development in London. She has a PhD in biodiversity management and over 10 years' experience working on climate change, with particular focus on how best to help those who are most vulnerable cope with its impacts.

CHAPTER 4

Rural livelihood diversification and adaptation to climate change

Terry Cannon

Most poor people in developing countries have livelihoods that are highly sensitive to climate change, because of dependence on farming, fishing or pastoralism, or as forest users. This high level of climate dependency means that adaptation to climate change is difficult if the people remain locked into such livelihoods. This chapter argues that a significant shift is needed that takes a large share of the population out of major reliance on climate, and into alternative rural livelihoods. This may be especially relevant in south Asia, where a large proportion of the rural poor are landless, and so unable to have much control over adaptation processes. Expansion of the rural non-farm economy (RNFE) may therefore also enable many people to sidestep the problems of poverty resulting from highly inequitable land tenure systems and their related power systems. It is suggested that adaptation funding may provide the basis for the required investment in rural infrastructure and other assets.

Keywords: climate change adaptation, alternative livelihoods, diversification, rural non-farm economy, labour, India, China

Much of the thinking and policy design on adaptation for the rural economy is defined in the context of continued (adapted) agriculture. This is inherently difficult, because of the following reasons:

- There is a very high extent of 'climate dependency' in such rural livelihoods, with most rural people directly affected by climate for most of their subsistence and earnings.
- It is extremely difficult to know exactly to which changes farmers need to adapt, since scaled-down climate models for localities do not exist, and may never be sufficiently valid at local levels.

Ideas for rural adaptation are framed around changes to farming, involving different varieties of crops (that are able to deal with higher temperature, drought, salinity and so on) and cropping systems. There is little scope for preparedness for new crop pests and diseases. What seems to be forgotten in this emphasis on adaptation through crops is that agricultural change in the past was seen as a potential route to poverty reduction through accumulation and a transformation away from subsistence. In the 1980s and 1990s, there was much more emphasis on agriculture providing pathways for rural growth.

http://dx.doi.org/10.3362/9781780447902.004

Instead, in the face of climate change, the focus seems to be more on substitute crops. Also in the past (from the 1980s to the 2000s), there was a very significant discussion of rural non-farm diversification of livelihoods – variously labelled the rural non-farm economy (RNFE), off-farm employment, income-generating activities, etc. This research was intended to both recognize and understand that a significant part of the rural economy was already 'non-farm', and that its growth would enable more rapid poverty reduction than a focus on farming alone (examples of such work include Ellis, 2000, and, for Africa, Bryceson and Jamal, 1997, Bryceson et al., 2000). It was also a result of an academic challenge to the idea that agriculture alone could enable a transition out of poverty.

This chapter explores the possible need for, and role of, the RNFE as a crucial part of adaptation. In part, it also asks questions about whether and how adaptation funding should have a role in supporting the emergence of the RNFE as a way to enable adaptation. It is partly a 'friendly critique' of community-based adaptation (CBA), on the grounds that CBA initiatives are currently too focused on adaptations to farming, and are not providing the basis for significant livelihood diversification at the local level. In addition, there is an inherent problem with CBA as currently conceived in that it is restricted to a limited number of communities where different agencies are involved (typically local or international non-governmental organizations (NGOs)). This will be far from enough to deal with the adaptation needs of everyone, which means having an approach to CBA that can take place *everywhere*, including in the majority of places that NGOs never can reach.

There is a wider context that is almost completely ignored in most CBA activities. For rural and agricultural adaptation, very little attention is being paid to the problem (especially in much of south Asia) of land tenure systems and power structures that determine access to land (and water).[1] As a result, there is potentially a serious gap in understanding about whether existing systems are viable (in terms of access to land, continuing livelihoods, current indebtedness and landlessness – the usual problems of rural poverty and development) under existing conditions, let alone with the effects of climate change. All these problems predate concern about climate change and adaptation, and will remain significant unless adaptation processes can sidestep or overcome existing power relations. If unequal land tenure and access to water – and the related issue of division of landholdings into small parcels – is one of the most fundamental problems for rural systems (and a key cause of poverty), why is it all but being ignored in the context of CBA?

These background 'development' problems call into question the validity (and fairness) of adaptation measures for rural economies that are already often in crisis. In other words, if existing problems of rural poverty and inequality are very hard to fix, what scope is there that this might get any better just because climate change has come along? Does adaptation funding offer an opportunity to combine adaptation to climate change with the right type of rural transformation? To put it another way, rural adaptation cannot be separated from dealing with existing rural development problems, since the causes of those problems are also highly likely to be barriers to successful adaptation, especially for poor people.

In this context, it is also worth noting the significant role that is being assigned to the 'community', without much of a critical discussion of the validity of this concept in relation to rural power relations. The term 'community-based' is put in front of many problems (for example, disaster risk reduction, health, water and sanitation): there seem to be few issues that it cannot fix. We therefore have very high expectations of 'Saint Community', who is supposedly able to perform miracles when anything is 'community-based'. While this issue cannot be discussed here in depth, it is crucial to be aware of two types of issues involved:

- First is the falsehood of community itself – in the sense that communities are never homogeneous, unified or inherently co-operative. This uncritical 'cuddly', 'fluffy' image is not deserved in many places: communities embody local (and wider) power relations, and there is no guarantee that grassroots works, even if we believe that grassroots is best (Brint, 2001; Cannon, 2008; Morcrette, 2009; Williams, 2003).
- Secondly, anything that is 'community-based' relies on using forms of participation, but there is a significant literature that challenges the validity of participatory methods. There are also earlier debates about problems of 'elite capture' of projects, and concerns about what happens in relation to local power systems after an NGO departs from the location. Does the project activity actually alter the power relations in a positive direction? (See, for example, Cooke and Kothari, 2001; Hickey and Mohan, 2004; Mohan and Stokke, 2000; Platteau and Abraham, 2002.)

Despite these problems, it is almost inevitable that adaptation needs to be supported at the local, 'community' level. But it is wise to ensure that the constraints are understood and acknowledged, and no pretence made that the root causes of problems can be evaded. From this awareness, there are two potential approaches for rural adaptation that need to be understood:

- The first is whether rural livelihoods can be transformed sufficiently to enable adaptation in situ (i.e. within the rural economy), and what the role of diversification to non-farming livelihoods can be in this process.
- The second is the need to recognize that, in some locations and circumstances, rural agriculture and its livelihoods may become impossible, requiring abandonment or retreat from some areas.[2]

This chapter focuses on the first of these issues (although it acknowledges that it may need to be linked to the second). It explores the existing knowledge on the RNFE as a way to reduce the climate dependency of significant numbers of (especially poor) rural people (Sabates-Wheeler et al., 2008). It is also essential to assess the potential for alternative organizational systems for access to land and the use of different forms of employment (e.g. to deal with the needs of landless peasants). This is examined generically, but current expansion of this work will look at Bangladesh and India more specifically.

What is the rural non-farm economy?

We know about the formal employment sector in towns and cities, but we have no difficulty in recognizing the 'informal sector'. This involves self-employment, micro, small and medium-scale enterprises (MSMEs) and trading activities – a spectrum of hundreds of types of good and not so good activities that provide livelihoods for millions of people. In the countryside, these types of activities blend in much more with the agricultural systems, and the rural economy tends to be looked at through an agricultural lens, as if farming is what is important and the rest of the activities are a residual category. The previous discussion and literature on the RNFE argue that this is wrong, and that both the *existing* and the *potential* non-farm activities must be taken much more seriously – mainly as a basis for poverty reduction. We now need to consider it as a possible basis for adaptation.

The RNFE can be defined as the array of 'livelihood activities (including employment and self-employment, formal and informal, legal and illegal) based in rural areas or pursued by people who are from households that are mainly rural-based, which do not involve direct agricultural production in crops or livestock' (NRI RNFE Project Team, 2000). Generally, primary sector production activities (crops, livestock, fisheries or forestry) are not included, but it is sometimes difficult to separate some of these, and people themselves are not likely to make such academic distinctions. There is no fixed separation between non-farm and farm, and many people may have a variety of livelihood undertakings. Within a household, there may be a mix of activities across the different household members. Clearly the ability (or desire) to engage in different types of activities – farm or non-farm – is strongly related to the asset portfolio (the bundle of different types of 'capital' – human, financial, social, physical or natural) of each individual and household.

A further definitional problem concerns what is *rural*: at what point does a settlement become a town, and therefore activities that go on in it are no longer considered rural? In my view, we do not need to be too concerned about this, in the sense that we know what we mean and we know what we want – to understand the difference from farming.[3] There is an issue about how small- and medium-sized towns *interact* with the countryside around them, and in this sense we need to understand what the types and quantities of the linkages are, and how they do (or do not) support or arise from the RNFE. In other words, do towns play a role in stimulating (or responding to) the non-farm activities going on around them? And if so, is there a possible role for investment (from adaptation funding) in such places as investment centres or 'growth poles' that can be targeted to provide support to the RNFE of their hinterland?[4]

So, the RNFE is highly relevant in the context of:

- absorption of rural surplus labour;
- spreading of risks for farm-based households;
- escape from agriculture (or diversification within farming households) to activities with higher returns;
- the need to subsist when farming has failed and alternatives become essential to survival and to coping.

RNFE activities may be:

- accumulative (e.g. setting up a small business);
- adaptive (e.g. switching from cash crop cultivation to commodity trading, perhaps in response to drought);
- coping (e.g. destroying natural capital, for example selling wood or making charcoal, in response to a shock);
- opportunistic (e.g. seeking non-agricultural wage labour as a higher or more stable income source).

Why is the rural non-farm economy of interest in the context of adaptation?

Hundreds of millions of rural people, many of them poor or near-poor (and therefore likely to become poor through the negative effects of climate change) are almost entirely dependent on climate for their farm- and natural resource-based livelihoods. Their context is that existing development-related problems are often not dealt with properly, even without having to adapt to climate change. In many parts of the world, the rural population exceeds the number of people needed to grow crops at existing levels of technology. This is largely an outcome of land tenure systems and power relations in the rural economy – systems that maintain these inequitable relations and force poor people to become indebted, suffer malnutrition and hardship, or migrate.

In relation to this, there is very little discussion in the adaptation literature or policy on what adaptation would mean for landless labourers (who are often estimated to comprise more than 40 per cent of households in much of south Asia) – or, for that matter, for any other rural economic and social (e.g. gendered) class as defined by share of control or ownership of land.[5] This is clearly crucial if the locus of effective adaptation is supposed to be at community level, given that for many countries (and especially in south Asia) the community consists of very unequal classes of landholders.[6] What adaptation might look like for landless farming households is difficult to say. We can therefore assess this problem through the entry point of the RNFE, in order to see if it can provide both its poverty reduction function (as espoused in the past) and an adaptation function for dealing with climate change. The issue, then, is whether the RNFE can enable forms of adaptation that reduce climate dependency, poverty and inequality and deal with different tenure groups (including females and males). Added to this is the question of whether adaptation funding has a role to play in supporting this adaptation function.

What is the rural non-farm economy and how significant is it?

The RNFE is heterogeneous, complex, varies over time and place, and responds fairly quickly to external factors (such as incentives from the market or policy).[7] It is often divided into *distress-driven* activities (which highlights the fact that RNFE activities are not always 'good' for people, the environment or development) and *opportunity-driven* activities. Distress-driven activities can include deforesting to make charcoal, sex work, begging and so on; these may merge conceptually with what elsewhere are called coping strategies. Opportunity-driven activities (which are chosen preferentially over existing activities including farming) involve income earning and/or investments that shift the dependence of the person or household away from a reliance on farming. This can include processing farm outputs (hence still climate dependent, but with the possibility of increased control over value addition), handicraft or artisan occupations (e.g. basket or pottery making, metal working or furniture making), manufacturing (as employee or employer) or trading.

Engagement in the RNFE is therefore largely determined by the pattern of assets in the hands of individuals or households, and their ability to make beneficial transitions to increase livelihood opportunities and avoid distress-driven activities (especially those that are environmentally or personally damaging). Clearly, this reflects the patterns of ownership and control of assets, especially land. Given that there are considerable constraints that may affect success for those with poor allocations of assets, it is important to understand those existing conditions while dealing with climate change. There are two key differences (and opportunities) compared with this existing situation: first is the chance to think anew because of the climate challenge, to develop policies that include livelihood diversification as a key component of adaptation. Second is the possibility that adaptation funding could provide the investment funds (and incentives) to make this possible.

Neither of these opportunities is guaranteed: despite promises about adaptation funding from 'Annex 1' (Organisation for Economic Co-operation and Development) countries, the amounts granted so far are well below the pledges, and some have been in the form of loans. There will also be disputes about whether or not support to livelihoods is a valid use of adaptation funds – and about whether it is different from 'development'. The first need will be for a lot more serious thought on the problems of proving the 'additionality' of support for livelihood diversification, and modifying past research on RNFE as a starting point for understanding what is needed and possible. As with other aspects of adaptation, we do not need to start from scratch: a great deal of work in development studies provides key knowledge that needs to be applied in the climate change context.

What do we already know about the RNFE? Generally speaking, the RNFE is already significant. One conceptual difficulty is that those who work on rural development issues tend to have an image of farming in their minds. This is now being carried over into the image of what has to be 'adapted' (i.e. crops). And yet the RNFE is estimated to occupy between 30 and 40 per cent of rural people in developing countries (Ellis, 1999; FAO, 1998). 'Occupy' does not mean full time (non-farm activities are often combined with farming), nor does it necessarily mean that they are adequate. Some of the data for RNFE activities is likely to include distress-driven activities that can hardly be considered beneficial or something to emulate. The activities differ in some cases between men and women (who tend to do less of them), and by age group (Ellis, 2000). Also, RNFE should not be seen as a single entity or sector – there are many types of activities, and it varies a lot between places and countries. In other words, policies and interventions are likely to need to be specific to circumstances.

Agriculture as a driver for non-farm activities?

A lot of the literature on RNFE considers that growth in agriculture is a key driver of its expansion. This is both because crops can provide inputs into processing and marketing, and because farm profits can provide investment. Expansion of farming also creates demand for farm inputs, machinery making and maintenance, along with ancillary services such as transport. Rising incomes from farm growth generate demand for things such as better houses, indoor plumbing and consumer goods – some of which may be produced locally and/or provide local employment. These backward and forward linkages may have a considerable impact on the rural locality (although, of course, some consumer and producer goods may be brought in from outside), creating a local multiplier effect in which money circulates locally, enables accumulation within the community, and builds up a virtuous circle involving a set of interdependent producers and consumers.

Some of the main examples of rapid expansion of RNFE have occurred in places where significant changes have happened to agriculture over a short period (e.g. parts of India and China), generating significant local multipliers and the expansion of villages into small towns, and even towns into cities. A key factor in these rapid expansions was the input of very significant public spending into the farm sector, which then generated spin-off effects in the RNFE. In other places, it has been argued that migrant remittances have had similar significant impacts on rural development as a form of cash injection (e.g. in Kerala). As will be seen, I am suggesting that this might be analogous to the potential for inputs from adaptation funding. The problem is that the preconditions of land tenure and rural power relations have a considerable impact on who gets access to the increased public spending on agriculture, and how the RNFE emerges; it is not simply a matter of a cash injection having the same effect in every location.

We therefore need to understand whether there are circumstances in which the RNFE can be stimulated in the absence of significant growth in agriculture. This is because raising output from farming may not be easy with the effects of climate change in many regions. So investment may have to feed directly into the RNFE, bypassing the stage of growth of farm output that has been assumed in the past as the stimulus. In the absence of other types of investment for this purpose, I am advocating that adaptation funding is explored as a means to deliberately foster the RNFE, as well as to support adaptation in agriculture. The justification for this is that removing people from direct and immediate climate dependency (out of farming and into non-farm activities) is a better form of adaptation than the uncertainty of continued farming. It may have the additional benefit of reducing some of the existing constraints on poverty reduction that arise out of land tenure systems that lock people into poverty.

Climate change and agricultural diversification

If growth of the RNFE requires prior profitable expansion of the farm sector, then clearly this has to take into account the effects of climate change on agriculture and its potential for expansion. An agricultural sector that cannot grow or is put into decline because of negative climate trends is not a good basis for livelihood diversification except into distress-driven activities. However, although, because of climate change, it may be essential for investment to go into the RNFE directly (bypassing any requirement for prior growth in agriculture), it is still necessary to explore diversification that is specific to farming itself.

The first steps in standard models of rural adaptation involve various types of crop changes and/or diversification that deals with trends in temperature and rainfall, anticipated shifts in seasonality and potential hazard shocks (noting that the forms of adaptation required for each of these may need to be different). Crop diversification can include:

- using new varieties of staple food crops (e.g. crops that are tolerant of higher temperatures);
- growing new food crops (farmers in many parts of the world have adopted maize as a less rain-dependent staple);
- shifting from subsistence to marketed crops (food and/or non-food) to increase income and purchase food instead of growing it all;
- moving more from (marketed) food crops to traded non-food crops (e.g. to cotton, jute or biofuel);
- changing from subsistence or marketed staple foods to traded non-staple foodstuffs (e.g. to coffee or fruit).

Key issues for those farmers who shift to greater selling rather than self-consumption include whether they can:

- obtain food on the market at an affordable price;

- get a good price for their marketed crops;
- retain more of the value addition (e.g. in processing crops) in the household or community.

These are not technical issues: there is the very significant socioeconomic context underlying the ability or desire to make such changes. In many parts of the world, while farmers are often willing to innovate and try new crops and production systems, there is caution about reducing direct access to subsistence crops and about reliance on having to buy food on the market. And all these shifts would be within the frame of local land tenure and power relations.

Factors affecting crop changes

The focus of this chapter is not so much concerned with agricultural adaptation. However, before moving to the RNFE, it is worth noting that many of the factors that are crucial to successful shifts in crops can be linked to non-farm activities, either upstream (inputs to the process) or downstream (produced from the crops), and so there are potential co-benefits and multiplier effects involved. A number of the key factors for crop changes are problematic, including the availability of seeds suitable for adaptation, other inputs, technical knowledge and training, and water availability. Underlying all these are also land tenure and the rural political economy and existing power relations. Clearly, some crop shifts could be maladaptive, for example in requiring more water or energy. It is entirely possible that, given existing power relations, some farmers will 'adapt' for benefits that avoid particular climate issues in the short term and that may not be viable or adaptive for the long term (e.g. unsustainable extraction of groundwater).

Adaptation in farming may (or may not) bring increased incomes (although we also need to analyse which economic and social groups benefit, and what happens to the landless). Assuming that incomes are raised, then this may produce investments for the RNFE. In most adaptation interventions, this potential to lead to diversification out of farming has not been very significant, and is a linkage that needs to be looked at much more. In a context of rural poverty, it is also likely that any rise in income for poorer people will lead to increased consumption rather than investment. We also need to assess the balance of effects on the work and time budgets of men, women and children, and on relations within households associated with that (and, for instance, any associated changes in assets, and likelihood of migration).

The rural non-farm economy: factors affecting non-crop diversification

The goal of earlier RNFE research has generally been to reduce poverty by supporting ways to extend livelihoods that give higher incomes outside agriculture. Given that there are already significant numbers of people engaged in distress-driven RNFE activities, the research has also assessed how to promote activities that can support households that have inadequate assets for proper subsistence and accumulation. By contrast, agricultural policies have often been aimed at increasing output, often with a food security focus, rather than reducing poverty. Poverty reduction has often been an indirect result, and in some cases inequality has increased and poverty has not decreased. In some cases (as we will see) agriculture policies have led to a significant expansion of RNFE activities as an unintended consequence of farming growth. In other places, this has not happened, suggesting that expansion of the RNFE is not an automatic outcome of agricultural growth, and needs to be treated as being context-specific.

One set of constraints is linked to (personal) funds available for investment, which, of course, is strongly related to economic and social class position, landholding and the ability to retain value at household level. Credit systems and an enabling business environment – a legal framework, property rights, protection from crime, adequate marketing, transport, skills and electricity supply – have been identified as key factors (Dasgupta et al., 2004; Davis and Bezemer, 2004; Ellis and Freeman, 2004; NRI RNFE Project Team, 2000). Many of these factors also relate to the role and significance of small and medium-sized towns, which have also been identified as important for providing nearby services, markets and integration (ibid.).[8]

What now needs to be explored is the potential role of adaptation funding as a support mechanism for diversification – not for farming (where its role is uncontroversial) but for the promotion of the RNFE. The methods and instruments for this will need to be designed, and could include credit systems, subsidies and social protection payments (adaptive social protection measures have already been discussed and developed in part at the Institute of Development Studies (IDS); see, for example, Davies et al., 2009). It may also be necessary to devise methods of cash payments that initiate a local multiplier that supports non-farm activities, or that enables farming to produce a surplus for investment. Hanlon et al. (2012) have argued the case for cash payments. An interesting Oxfam project in Vietnam experimented with cash grants (Oxfam, 2008), after which Oxfam tracked what happened: the results suggest there are considerable benefits (though this project was not related to adaptation). It is clear that in order for adaptation to succeed, very new thinking on alternatives to conventional actions are needed, and that adaptation funding may enable or be the basis for this. However, an evidence base will be needed to justify such investments, and this makes community-based action research (such as the Action Research for Community Adaptation in Bangladesh project) imperative as the beginnings of that process of providing justification (van Aalst et al., 2008).[9]

Which areas have seen the rural non-farm economy expand rapidly?

This chapter does not provide a global survey of examples, but instead looks at two cases that involve very different contexts and socioeconomic outcomes. They underline the great importance of context – especially land tenure – as a factor in the success of RNFE policies in adaptation. The case studies are of parts of India (in the 1960s to 1970s during the Green Revolution) and of China (after the start of the rural economic reforms and responsibility system in 1983).

China and the impact of the rural economic reforms after 1983

The economic reforms that began in 1979 led China to the most rapid and lengthy period of economic growth of any country in history. For almost all the past 30 years, growth has been above 9 per cent per annum. A significant share of that occurred in rural enterprises (in the industry and service sectors), whose rate of growth was much higher than that in the urban and state-owned sectors in the first decades. In China, these rural and quasi-rural enterprises are labelled 'township and village enterprises' (TVEs) – a term that indicates that in many cases they are based in small towns (some of which became large towns or even cities as a result of that growth). By 1996 there were over 135 million people working in TVEs, although this was probably the peak year for such employment (it had dropped to 127 million by 1999). By the mid-1990s, TVEs were accounting for over 30 per cent of the country's gross value of industrial output, with individual and private enterprises (many of them in rural areas) accounting for a further 25 per cent. In other words, after 15 years of reforms, rural industry had outstripped the state-owned enterprises, which were producing only 34 per cent of total output. Rural enterprises were producing 17 per cent of electronics goods, 26 per cent of machinery, 40 per cent of coal, 80 per cent of garments, and 95 per cent of small and medium farm tools. Moreover, rural industries generated 36 per cent of total export earnings by 1996 (Smyth, 1998: 784; Cannon 2000; 2004).

We must not omit the non-industrial rural enterprises, as these are also significant in terms of employment and output value, although they tend to grow at a slower rate ('only' 20 per cent per annum, compared with an average of 25 per cent for rural industry in the period 1978 to 1988). There are three other major rural non-farm sectors – construction, transport and commerce – and by 1988 these accounted for a quarter of the output value.

There were several factors that led to this growth in rural industry and enterprise (which includes very significant growth in the service sector). It is also important to see how these factors *interacted*: it was their mutual reinforcing of an environment conducive to growth that was important. It is also vital to note that the factors include a *very significant role for state intervention*: despite the wishful thinking of neoliberal ideologues and proponents of market forces, China's growth has been in large part a consequence of government intervention rather than only of markets and 'capitalism'. One of the key factors in the rapid growth of rural enterprises was a sudden large rise in state procurement prices for key crops, linked to the de-collectivization of farming (in the 'responsibility system') around 1982, which gave rise to private production incentives for farmers. After selling the required procured outputs to the state (at the new higher prices), households were allowed to retain profits from sales of any surplus produce. Soon after this, the previously rigid prohibition of rural–urban migration was relaxed, enabling surplus rural labour (usually one member of a household) to move to jobs in other parts of their province or in distant provinces (including to jobs in the rapidly expanding industries in parts of the east and south coast). Remittances were also used by some enterprising households to invest in home village enterprises.

In all of this, a crucial role was played by local governments, especially at township (ex-commune) level and county level (and even down to village level). These local governments acted as entrepreneurial 'holding companies' with both the property rights that enabled them to control and benefit from businesses, and significant incentives (created by central government policies) that encouraged them to initiate new industry and service sector activities. However, not all regions were able to succeed in this way, and while many local governments in the coastal provinces were able to thrive, others in the interior could not initiate activities or compete. In part this was because success was greatest where all the factors were present and interacted.[10]

Conditions in China were clearly very different from those in any other part of the world, so it is doubtful that any of this can be replicated elsewhere. But the point of the example is to examine a few key factors that may be relevant in how it happened. In 1980–81, the government abolished the People's Communes and initiated the responsibility system. Land was shared out between all rural households, which were allowed to retain or sell their output over and above a fixed procurement of key crops for the state. This was a very powerful incentive for farmers to increase their output of both procured crops (those that had to be sold to the government at a price set by the government) and other crops that could now easily be sold on the market. Almost simultaneously, the government made a significant increase in the price it paid to farmers for key crops – a 50 per cent increase in the price for grain, and a 30 per cent increase for cotton. Peasant output soared to capture this price increase, and farmers' incomes went up enormously over a very short period. This crucial factor of increased procurement prices is often forgotten in the literature on the success of China's TVEs.

This improvement in the agricultural terms of trade lasted until about 1985, by which time the price advantage was eroded. But with this significant increased purchasing power in the rural economy, people began to demand higher-order foods (meat, eggs and vegetables), and this contributed to rising farm incomes. Rising incomes in the cities and towns also led to a rapid increase in demand for such foods, which could be sold on the free market. And with higher incomes, rural household demand for improved or new houses and consumer goods (some made locally, such as furniture) increased significantly.

Thus the initial stimulus for the growth of TVEs was a significant rise in government spending, which led to the increased purchasing power and investment potential of the rural population; and the spending seems to have had very high local multipliers. One of the most significant areas of spending was in house improvement and construction, which used a lot of local materials and labour. Farmers also increased their use of some farm inputs, leading to increased trading. (However, the newly subdivided plots were far too small for any mechanization, and irrigation systems constructed under the collectives fell into disuse and disrepair.) Inputs and outputs also required transport, and significant growth in the transport sector.

Some people were also encouraged to become private entrepreneurs, starting businesses in manufacturing and services (especially road transport, restaurants, retailing, trading and construction, among others). These entrepreneurs included peasants who had surplus capital, the right connections and openings. They also included local government and communist party officials who were able to use their official position to protect their business activity. Migrant remittances (and returned migrants with savings and new skills and acumen) also contributed to rural growth.

Local governments, property rights and incentives. When the central government abolished the communes, it did not abolish the power of local governments to 'own' and control local enterprises. In some parts of the country (see above) these included quite significant rural industries. But even where there were no significant pre-existing rural enterprises, local governments were in effect inheriting the property rights of a collectivized system where local government 'owned' and controlled local enterprises. So both pre-existing and any new enterprises were effectively under local government control. This even applied to private enterprises, since it was normally difficult for a private businessperson to operate without the enabling environment of government.

Local governments down to the village level therefore had a powerful incentive under the reforms to engage in business activities, including manufacturing and services, because they had a stake in the profits and other benefits (status, promotion and fringe benefits – including cars, banquets and travel) that arose from the activities they could promote. However, this incentive, which was derived from property rights, was only likely to produce results in conjunction with other policies, such as the fiscal (and foreign exchange) regime, that provided the real incentives for investment.

Here then are some key factors that *may* provide some guidance for the emergence of the RNFE in other countries: a sudden and significant influx of public funding (in this case a top-up to the procurement prices); a farming population with the incentive and ability (because of landholdings being relatively equal) to increase output; and a local government framework that promoted and shared in the system. Note that while a very large number of the rural population moved out of farming into RNFE activities, food production went up (although there were negative outcomes such as the abandonment of irrigation systems and mechanized farming, which became impossible on the smaller decollectivized plots).

India's rural development in the Green Revolution compared with China

The closest parallel with China that might be found in India is the rural boom that accompanied the Green Revolution in Punjab, Haryana and parts of west Uttar Pradesh from the late 1960s to the late 1970s. As in China, the initial impetus for the growth was a significant increase in procurement prices: in India this was for staple grains (mainly rice and wheat) that were being produced with the new high-yield variety (HYV) methods in water-secure districts. The fairly sudden increase in farm incomes in Green Revolution areas also had large multiplier effects, in house construction, domestic durable consumer goods, and also in farm inputs, including the mechanization of fieldwork and much increased irrigation.

Surplus capital also became available for investment in new manufacturing enterprises, and in the service sector. The streets of the growing towns (some of which had been villages not long before) of the Green Revolution districts were alive with trading, manufacturing activities and repairs going on along the roadside. Towns such as Ludhiana and Jalandhar (Jullunder) experienced rapid growth as service centres and investment outlets for the minority new rich local farmers.

In China, the rural expansion of TVEs was a result of the disbanding of collectives, which meant the distribution of equitable shares of land to all rural households. By contrast, the Indian experience of the Green Revolution was of a change in technology that was overlaid on already inequitable land tenure, an inequality that was generally widened by the process of change. The inequality was further reinforced through state subsidies to large farmers for agricultural inputs: fertilizer and pesticides were subsidized, electricity and diesel fuel (for irrigation pumping) were cheap, and water was free. This meant that richer groups generally experienced a disproportionate increase in income and savings. Poorer farmers and the landless generally benefited only from increases in employment (with in-migration), without the significant rise in disposable income experienced in China among a large proportion of the rural population.

This crucial difference remains valid today, and so a policy that brought higher incomes to farmers in India would not have the same effect on income distribution, disposable income and rural industry multipliers as happened in China. In any case, there is much less scope for any policy intervention in India through procurement price increases: in the post-1990 liberalized economy, the government's capacity to act in regard to commodity prices is much more limited than it was in the 1970s. The central government has also had chronic budget deficits, and this is taken to rule out significant increases in public expenditure.

However, in both the case of India's Green Revolution and China's economic reforms, it is evident that the rural 'take-off' of non-farm activities would not have happened without a significant transfer of public funds. Would other policies (not related to crop prices or input subsidies) bring similar results for the purposes of generating support for the RNFE as part of adaptation? In India, the increased farm income fuelled the industrial revolution that accompanied the Green Revolution. In China, it provided the initial impetus, which was then sustained by local spending and reinvestment encouraged by local governments under their fiscal contracts with central government (in which they were permitted to retain a significant amount of locally generated revenue).

Moreover, and this is highly significant, much of the value addition took place in the rural localities; rural growth in both countries occurred because much of the spending and investment was carried out in the locality where incomes increased, and led to local entrepreneurship being stimulated for both production and consumption goods. In both countries, significant targeted increases in government spending (through what are now inappropriate or impossible means) were quite crucial in stimulating the rapid growth of the RNFE, even though this was not the initial aim of the policy.

In other words, the rapid growth of non-farm livelihoods was not a result of specific policies to promote the RNFE. Instead, it grew as a side effect of agricultural policies. It is doubtful whether there has ever been anywhere a similar degree of success in promoting the RNFE by means of policies that were simply directed at those activities in isolation. This is a crucial policy issue: by comparison, how successful can policies be that are specifically aimed at promoting the RNFE? And how likely is the success of any policy that does not involve an initial, fairly significant, investment? And could that investment come from adaptation funding?

But there is an important difference between the two countries that needs to be understood for the context of adaptation funding. Not only do most countries (especially in south Asia) have no history of equitable land distribution or collective ownership of rural enterprises, they also have no substantial history of local governments being involved in entrepreneurial activity or acting as significant economic players. In China, the enterprise economy that emerged could not have happened without the 'entangled' local governments, which promoted, managed, initiated and fostered many of the non-farm activities. So if we consider this to be a crucial factor that led to China's RNFE growth, does that give any insights into what might be relevant elsewhere, and might it be something that could be the role of local government in relation to adaptation funding?

China's experience suggests two key issues to address: one is local governance and the way in which an environment conducive to beneficial state (and private) investment can be created. The other is the imperative that the government act as 'entrepreneur of last resort' and makes investments (not subsidies) that support economic activities and income growth for adaptation in the countryside. The major difficulty is in finding investments that contribute very quickly to income growth in order to generate multipliers, with backward and forward linkages, since it would be pointless simply making investments in activities with outputs for which there is as yet no effective demand.

Adaptation funding, the rural non-farm economy and community-based change

In both India and China, we know that increased spending stimulated demand for local products and services. Much of this demand was satisfied by local entrepreneurs who responded to new opportunities and expanded the RNFE. However, an initial public investment (cash flows as subsidies or increased crop prices) was essential to start the process. Rising income in farm households was able to stimulate purchases of producer and consumer goods, to increase the forward and backward linkages between farms and their customers and suppliers, and to generate high levels of local value addition and strong multipliers. In both India and China, the increased production became self-sustaining, and created increased wealth and government revenue (along with significant environmental problems that cannot be discussed here).

State investments can be justified if they can become part of such a virtuous cycle, in which the government receives an equivalent or greater amount than it spends, through increased revenues from taxes on individuals and new enterprises. It is also imperative to ensure that they do not reinforce current unequal income and asset distributions (which is what happened in India), and offer genuine opportunities for poverty reduction (which many claim did happen in China). Would such investments be the basis for rural adaptation to climate change?

Who are the potential actors in the countryside who might respond to incentives to invest locally? In other countries there is no equivalent of the township and village governments that played such a significant role in rural growth in some parts of China. Who were the entrepreneurial actors in the Green Revolution or, more recently, in other parts of India where there has been significant rural growth (such as parts of Gujarat, Kerala and Tamil Nadu)? Can they be brought into partnership with Panchayat (local government) institutions so as to benefit local economic development? Is there scope for forming rural enterprise holding companies based in small towns, in which there is a partnership between entrepreneurs, local governments, community-based organizations (CBOs) and credit institutions to support adaptation?

These holding companies could involve fiscal contracts similar to those that were used in China: an agreement between entrepreneurs and local government over revenue sharing, and a contract between the local level and the higher administrative level for the amount of taxes remitted upwards (which eventually compensate central government for its public investment). The holding company would operate on the basis of transparency and supervision by CBOs, which are in effect its shareholders. This would act as a check on corruption and inefficiency.

Clearly, a key issue is that the labour available for farming in many parts of the world is greater than that needed for crop production (at a given level of technology). So the issue is how to solve this structural problem that gives rise to rural surplus labour, poverty and high levels of climate dependency, which will cause even greater hardships with climate change. These issues are already very difficult to resolve, and are deeply embedded in rural (and often national) power relations. In a number of countries, land reform measures that have been tried over the past 50 years have largely failed to reduce unequal holdings or landlessness. This problem will be solved by either policy or protest (or both, or by the latter forcing the former). But whatever the situation, the problem is going to become more severe with climate change. Interventions that aim to support adaptation without dealing with existing development problems, inequality and unequal land tenure are not going to succeed. One possible way out may be the promotion of the RNFE, using adaptation funding as the stimulus that supports both farming and non-farming activities. Growth of the RNFE may be a way to sidestep the rural power issues that at the moment maintain inequality and poverty. We have but a few years to get this right, and there are not many options to choose from.

My concern is that adaptation perceived through the lens of CBA is inherently narrow and restricted, and is predominantly seen as something that is implemented by organizations that cannot possibly cover every community in the developing world. For adaptation to succeed there must be relevant top-down policies that can support adaptation for everybody. I am not convinced that CBA as currently debated and practised can achieve that. Local-level work can provide evidence for what needs to happen in terms of changed incentives, adaptation activities, and deciding how adaptation funding can be used effectively to support this. But this is not nearly enough, and evidence gathered through CBA must be used to form top-down policies that are relevant for everyone and, as argued here, to shift many people out of climate dependency through the RNFE.

Policies can include social protection measures (e.g. conditional payments that move people towards adaptation), insurance systems, grants for training in new livelihoods for the RNFE, cash payments to boost local economies and stimulate multiplier effects, and investments into small towns as service centres and markets for rural areas. Nothing of what I am suggesting is easy, but then dealing with climate change is perhaps the greatest challenge ever faced by humanity, and we need new and big ideas. The focus of CBOs and NGOs on the locality and community will not work, since there are millions of villages they can never cover.

What happens for CBA must work without necessarily having an outside agency or local organization available to make it happen. Rapid economic and social change is clearly possible – as shown by the extraordinarily quick shift in some developing countries, such as Taiwan and South Korea. Other countries, or regions within some countries, have become much less dependent on farming as their main livelihood. These changes were produced through policies and new incentives, often combined with significant government funding (as in China and India) and investment (in Taiwan and South Korea for Cold War political reasons). The (adaptation) funds to enable such investments exist and will increase, although at the moment there is very little evidence on how best to use them. The work in areas where CBA is carried out is the way to provide this evidence, which is then used to design top-down adaptation policies that can work quickly. This will require taking risks and making experiments, as there is not enough time to ensure that we get everything right from the start.

Acknowledgements

The work related to this chapter, including a field visit to Bangladesh in April 2012, was partly supported by the Arkleton Trust and by the ARCAB (Action Research for Community Adaptation in Bangladesh) research project, which is funded by the Department for International Development. It has also been supported by the Climate and Development Knowledge Network project 'Getting Climate Smart for Disasters' in India, with IDS in partnership with Intercooperation and the All India Disaster Mitigation Institute.

Notes

1 This is probably because the main funders for and agencies of adaptation are institutions that have to function in a politically neutral context, unable to criticize (or even openly discuss) the significant constraints on rural development or poverty reduction that arise because of power relations. In my view, it is imperative to at least understand those constraints and speak out about them, rather than pretending that they do not have an effect.

2 This requires exploration of ideas of supported migration, in which adaptation investments are made into support for new livelihoods (available for both the migrants and the host communities), for example in small and medium-sized towns. Separately, I am exploring the reopening of the literature on 'growth poles' for this, and the scope for using adaptation funding to support such investments.

3 An interesting issue is that in some parts of the world where RNFE seems to have developed widely there is considerable difficulty in demarcating the boundary between town and countryside (as in parts of Kerala), or there is a 'rur-urban' quality to the settlements, as in parts of Indonesia where the term *desakota* has been given to the indeterminate amalgamation of town with countryside (see McGee, 1991; there is much recent work on using the concept in Asia and other continents).

4 The concept of growth poles was important in geographical analysis from the 1950s, and is still being used in some situations (e.g. World Bank report on Mozambique, 2010). It would be useful to explore their potential as the basis for supported migration to help people whose livelihoods are likely to come to an end because of climate change.

5 A literature is emerging on the relationship between land tenure and climate change, but tends not to be about how adaptation will take place for different classes of landholders. Examples of this literature include FAO

(2010) and Quan and Dyer (2008). A special issue of the open access FAO *Land Tenure Journal* has also appeared (Arial et al., 2011). Other work on land tenure concerns the links with mitigation (i.e. the role of land tenure in relation to mitigation in, for example, forests and carbon sinks).

6 I use the term *landholder* deliberately so as to include those who own land as well as those who have access to it as renters, sharecroppers or wage labourers.

7 It is sometimes referred to as the RNF *sector* – but since it is not a sector in the same sense as agriculture or industry, that term is not accepted here.

8 Two major research projects examined the RNFE in Africa and south Asia in the early 2000s: the RNFE project at the Natural Resources Institute (University of Greenwich), and the LADDER project at the University of East Anglia. The project reports are available at www.nri.org/projects/rnfe/papers.htm and www.uea.ac.uk/international-development/People/staffresearch/ladder.

9 Action Research for Community Adaptation in Bangladesh is a project involving around 10 international NGOs for testing adaptation measures in field sites in different parts of the country, to assess what can work.

10 For more details on these issues, see Ho (1995), World Bank (1999) and Yao (2002).

References

Arial, A., Hui Lau, T., and Runsten, L. (2011) *Land Tenure Journal*: Thematic issue on land tenure and climate change, 2(2011) <www.fao.org/nr/tenure/land-tenure-journal/index.php/LTJ/issue/view/4>.

Brint, S. (2001) '"Gemeinschaft" revisited: a critique and reconstruction of the community concept', *Sociological Theory* 19(1): 1–23 <http://dx.doi.org/10.1111/0735-2751.00125>.

Bryceson, D. and Jamal, V. (eds) (1997) *Farewell to Farms: De-Agrarianization and Employment in Africa*, Aldershot: Ashgate.

Bryceson, D., Kay, C. and Mooij, J. (eds) (2000) *Disappearing Peasantries? Rural Labour in Africa, Asia and Latin America*, Rugby: Practical Action Publishing.

Cannon, T. (2000) 'Introduction: the management of space and nature', in Cannon, T. (ed.), *China's Economic Growth: Impact on Regions, Migration and Environment*, Basingstoke: Palgrave Macmillan.

Cannon, T. (2004) *China and the Spatial Economy of the Reforms: Understanding Topocracy and the 'Local Developmental State'*, Occasional Paper Series No. 48, Hong Kong: Hong Kong Baptist University, Centre for China Urban and Regional Studies, Department of Geography.

Cannon, T. (2008) *Reducing People's Vulnerability to Natural Hazards: Communities and resilience*, WIDER Research Paper 2008/34, Helsinki: United Nations University <www.wider.unu.edu/publications/working-papers/research-papers/2008/en_GB/rp2008-34/> [accessed 9 November 2013].

Cooke, B. and Kothari, U. (eds) (2001) *Participation: The New Tyranny?*, London: Zed Books.

Dasgupta, N., Kleih, U., Marter, A. and Wandschneider, T. (2004) *Policy Initiatives for Strengthening Rural Economic Development in India: Case Studies from Madhya*

Pradesh and Orissa, London: Natural Resources Institute, University of Greenwich <www.nri.org/projects/rnfe/papers.htm> [accessed 4 October 2013].

Davies, M., Guenther, B., Leavy, J., Mitchell, T. and Tanner, T. (2009) *Climate Change Adaptation, Disaster Risk Reduction and Social Protection: Complementary Roles in Agriculture and Rural Growth?*, IDS Working Paper 320, Brighton: Institute of Development Studies (IDS).

Davis, J.R. and Bezemer, D. (2004) *The Development of the Rural Non-Farm Economy in Developing Countries and Transition Economies: Key Emerging and Conceptual Issues*, Chatham: Natural Resources Institute.

Ellis, F. (1999) *Rural Livelihood Diversity in Developing Countries: Evidence and Policy Implications*, Natural Resource Perspectives No. 40 (April), London: Overseas Development Institute.

Ellis, F. (2000) *Rural Livelihoods and Diversity in Developing Countries*, Oxford: Oxford University Press.

Ellis, F. and Freeman, H.A. (2004) *Rural Livelihoods and Poverty Reduction Policies*, London: Routledge.

FAO (1998) *The State of Food and Agriculture 1998. Part III: Rural Non-Farm Income in Developing Countries*, Rome: Food and Agriculture Organization of the United Nations (FAO).

FAO (2010) *Land Tenure and Natural Disasters: Addressing Land Tenure in Countries Prone to Natural Disasters*, Rome: FAO <www.fao.org/climatechange/65622/en> [accessed 4 October 2013].

Hanlon, J., Barrientos, A. and Hulme, D. (2012) *Just Give Money to the Poor: The Development Revolution from the Global South*, Boulder, CO: Kumarian Press.

Hickey, S. and Mohan, G. (eds) (2004) *Participation: From Tyranny to Transformation?*, London: Zed Books.

Ho, S.P.S. (1995) 'Rural non-agricultural development in post-reform China: growth, development patterns and issues', *Pacific Affairs* 68(3): 360–92 <www.jstor.org/stable/2761130>.

McGee, T.G. (1991) 'Emergence of desakota regions in Asia: expanding a hypothesis', in Ginsburg, N., Koppel, B. and McGee, T.G. (eds), *The Extended Metropolis: Settlement Transitions in Asia*, Honolulu: University of Hawaii Press.

Mohan, G. and Stokke, K. (2000) 'Participatory development and empowerment: the dangers of localism', *Third World Quarterly* 21(2): 247–68 <http://dx.doi.org/10.1080/01436590050004346>.

Morcrette, A. (2009) 'Critical about community: avoiding "stylised facts" in disaster vulnerability analysis', unpublished MSc thesis, London: London School of Economics and Political Science.

NRI RNFE Project Team (2000) *Policy and Research on the Rural Non-Farm Economy: A Review of Conceptual, Methodological and Practical Issues*, London: Natural Resources Institute (NRI), University of Greenwich.

Oxfam (2008) *Evaluation of the Cash Transfers for Development Project in Vietnam: Full Report*, Oxford: Oxfam GB.

Platteau, J.P. and Abraham, A. (2002) 'Participatory development in the presence of endogenous community imperfections', *Journal of Development Studies* 39(2): 104–36 <http://dx.doi.org/10.1080/00220380412331322771>.

Quan, J. and Dyer, N. (2008) 'Climate change and land tenure: the implications of climate change for land tenure and land policy', Land Tenure Working Paper 2, Rome: FAO.

Sabates-Wheeler, R., Mitchell, T. and Ellis, F. (2008) 'Avoiding repetition: time for CBA to engage with livelihoods literature?', *IDS Bulletin* 39(4): 53–9 <http://dx.doi.org/10.1111/j.1759-5436.2008.tb00476.x>.

Smyth, R. (1998) 'Recent developments in rural enterprise reform in China: achievements, problems and prospects', *Asian Survey* 38(8): 784 <www.jstor.org/stable/2645583>.

van Aalst, M., Cannon, T. and Burton, I. (2008) 'Community level adaptation to climate change: the potential role of participatory community risk assessment', *Global Environmental Change* 18(1): 165–79 <http://dx.doi.org/10.1016/j.gloenvcha.2007.06.002>.

Williams, A. (2003) *The Viability of Integrating Community Based Disaster Management within NGO Strategic Management*, mimeo, Coventry: Coventry University.

World Bank (1999) *Accelerating China's Rural Transformation*, Washington, DC: World Bank.

World Bank (2010) *Prospects for Growth Poles in Mozambique*, Washington, DC: World Bank.

Yao, S. (2002) 'China's rural economy in the first decade of the 21st century: problems and growth constraints', *China Economic Review* 13: 354–60.

About the author

Terry Cannon is a Senior Research Fellow at the Institute of Development Studies in the UK, where his main work relates to climate change adaptation, especially community-based adaptation, and also to vulnerability to natural hazards and disaster risk reduction. He also teaches on climate change and disaster risk reduction as a visiting lecturer at King's College London, the Universities of Copenhagen and Geneva, and at the International Centre for Climate Change and Development (ICCCAD) in Bangladesh.

Part Two
Case studies

CHAPTER 5

Assessing local adaptive capacity to climate change: conceptual framework and field validation

Alejandro C. Imbach and Priscila F. Prado Beltrán

Adaptation is a local-level process where the key stakeholders are the local population and their organizations. Assessing local adaptive capacity is a fundamental step in developing a local strategy for adaptation to climate change. This chapter presents a methodology for assessing local adaptive capacity based on the community capitals framework, which is useful to detect different constraints based on the different community capitals (cultural, human, social, political and financial) and to define specific actions to address those constraints. The chapter presents the results of its validation in two communities in Chiapas, southern Mexico. The validation showed that the framework was effective in identifying and disaggregating the key components of local adaptive capacity. The results from this analysis also led to a clear identification of different and specific strategies to deal with the identified constraints.

Keywords: adaptive capacity, climate change, exposure, sensitivity, perception, assessment, Mexico

The Intergovernmental Panel on Climate Change (IPCC) defines adaptive capacity as the ability or potential of a system to respond successfully to climate variability and change, and includes adjustments both in behaviour and in resources and technologies. In practice, this definition leads to the notion that adaptation is a local-level process where the key stakeholders are the local population and their organizations. How local communities respond to pressures for change is not restricted to climate issues; on the contrary, local communities are continuously subjected to pressure for changes ranging from changing market demands, technological changes and infrastructure development to changing access to credit, as well as many others. Preliminary work at this level has shown that the potential and constraints for local adaptation change dramatically from community to community, depending on a large number of different factors (cultural, human, financial, social, etc.).

There is a good body of work and experience in using the livelihoods approach to analyse complex situations. This work has recently evolved into the community capitals framework (CCF, see Flora et al., 2004), integrated

http://dx.doi.org/10.3362/9781780447902.005

by Imbach (2012) into a larger scheme (life strategies or '*estrategias de vida*'), linking what local communities do with their resources or capitals and what they get from those activities to satisfy their fundamental human needs and achieve a certain level of sustainable development. Based on this work, the authors decided to develop a preliminary framework of analysis for local adaptive capacity to climate change and to test it in rural communities in the Soconusco region of the state of Chiapas in south-western Mexico, bordering Guatemala.

The analytical framework

The framework ties the CCF community capitals (Flora et al., 2004) or resources (human, cultural, social, political, financial and infrastructure; Imbach, 2012) to the usual barriers or bottlenecks identified when dealing with change issues in community development.

A sketch of the framework shows the sequence of the analysis.

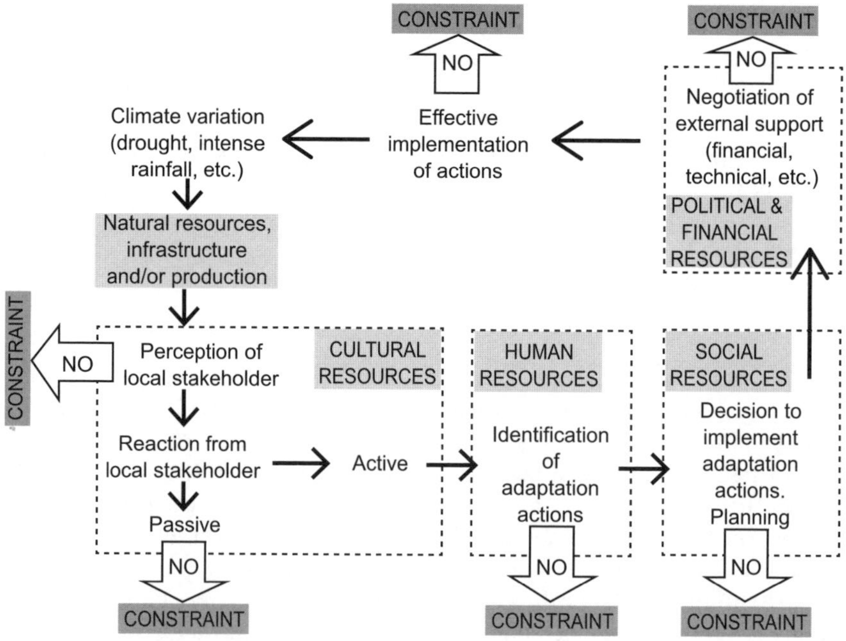

Figure 5.1 Adaptive capacity assessment framework

The framework is structured according to the following logic, and it is important to be aware of how each step either leads to another, different aspect to be considered or diverges from the path and requires action at that particular point.

1. *Identification of the exposure.* In other words, how does climate change or variability in the area manifest itself (including indirect effects such as floods, landslides) and what is its intensity?
2. *Identification of the sensitivity of the area to those changes.* How are local livelihoods and infrastructure affected?
3. *Perception.* Are these changes and effects perceived by the local population? Are they perceived as a changing trend or simply as part of the usual variable conditions prevalent in the area? Obviously, different perceptions of changes and effects should lead to different interventions.
4. *Reaction.* What is the reaction of the local people to the perceived changes? Do they think that something can be done or not? We think that this reaction is heavily influenced by cultural characteristics such as the way in which communities explain the natural world, its changes, the drivers of those changes and the role of human beings in the entire scheme. The basic notion about what they believe can be changed and what they believe cannot be changed is part of this area of enquiry. The reaction to the perceived changes and their effects can be passive (i.e. nothing can be done) or active.
5. *Identification of actions.* For those perceiving the changes and deciding to act, the next step is finding out what to do. This step is related to knowledge and experience, aspects that in the framework are linked to human capital and resources. At this stage, different levels of knowledge and understanding should be expected, leading to different types of possible action ranging from taking effective adaptation measures to leaving the region (emigration). There are other external factors influencing this stage, such as accessibility, market demand and the potential for alternative sources of income, but they can be incorporated into adaptive action schemes only if they have been considered alongside local knowledge.
6. *Preparation to implement actions.* The implementation of actions at the local level usually has initial requirements of local organization and basic planning, and both these aspects are linked to social capital or resources. It is true that communities also need funding and technical support (addressed at the next step), but these can be accessed only (in most cases) through already existing local organizations and basic planning (action plans or project proposals).
7. *Implementation.* This is the next stage and involves the negotiation of financial resources and technical assistance. This stage centres on political capital and resources. Local organizations and groups need to have the contacts and the capacity for lobbying for the allocation of support from external sources (including the government, non-governmental organizations (NGOs) and the private sector) for their adaptation plans and projects.
8. *Effective implementation and sustainability of actions.* This is the final step of the process, and the last bottleneck, and it consists of the capacity to actually implement what is planned in an effective and efficient manner and to be able to maintain the actions, infrastructure and other adaptive measures over time. In general, it seems that maintaining what was built or established

is harder than building or establishing it because of the erosive forces that emerge after the initial enthusiasm and momentum. This is a lesson learned from other community interventions in the past, such as infrastructure for soil conservation (terraces, barriers, etc.), which are also often relevant adaptive measures required by climate change and variability.

9. *Impact on climate change and variability effects*. At this stage it should become evident that the actions taken are generating some positive changes in livelihoods and resilience. This is a key factor in revitalizing any cycles of the adaptation process that are still progressing through the earlier stages, and allows those new cycles to benefit from successful experiences and lessons learned (both positive and negative).

Analysis of the proposed framework leads quickly to the idea that there are several bottlenecks and bifurcations along the route. As a consequence, it is to be expected that local populations will choose to take different paths to different extents, depending on their particular situation. Moreover, each one of the different groups that splits off along the way may require specific interventions depending on the causes of that split; groups that do not perceive impacts should be addressed differently from groups lacking planning capacity, and communities blocked by a lack of organization have different needs from those who do not believe that adaptation is possible.

Field implementation and validation

The field implementation and validation of the framework described above were carried out during 2010 by Prado (2011) in the Soconusco region of the state of Chiapas in south-western Mexico, as shown in Figure 5.2. Two rural communities (*ejidos*, a form of communal rural property specific to Mexico) were selected: Azteca and Manuel Lazos.

Azteca is small *ejido* (46 families) on the high slopes of the Tacaná volcano and can be considered a relatively isolated community, far from medium or large urban centres. Manuel Lazos is a larger community (450 families), also located on the Tacaná volcano slopes but very close to Cacahoatán, a medium-sized town in this predominantly coffee-growing area.

Both communities are located within the Cahoacán river watershed. This river has its source on the slopes of the Tacaná volcano in the Sierra Madre de Chiapas and flows towards the Pacific Ocean. The watershed is geographically located in the American tropics.

These communities are situated in the high (Azteca) and middle (Manuel Lazos) basin, in areas of steep slopes at altitudes between 400 m and 1,600 m. The average annual rainfall is 3,200 mm, concentrated between April and November with a dry season between December and March. The average temperature is 25 degrees Celsius, with an average maximum of 31 degrees and an average minimum of 19 degrees. The natural vegetation is mostly mixed broadleaf rainforests, but most of it has been converted to agricultural use.

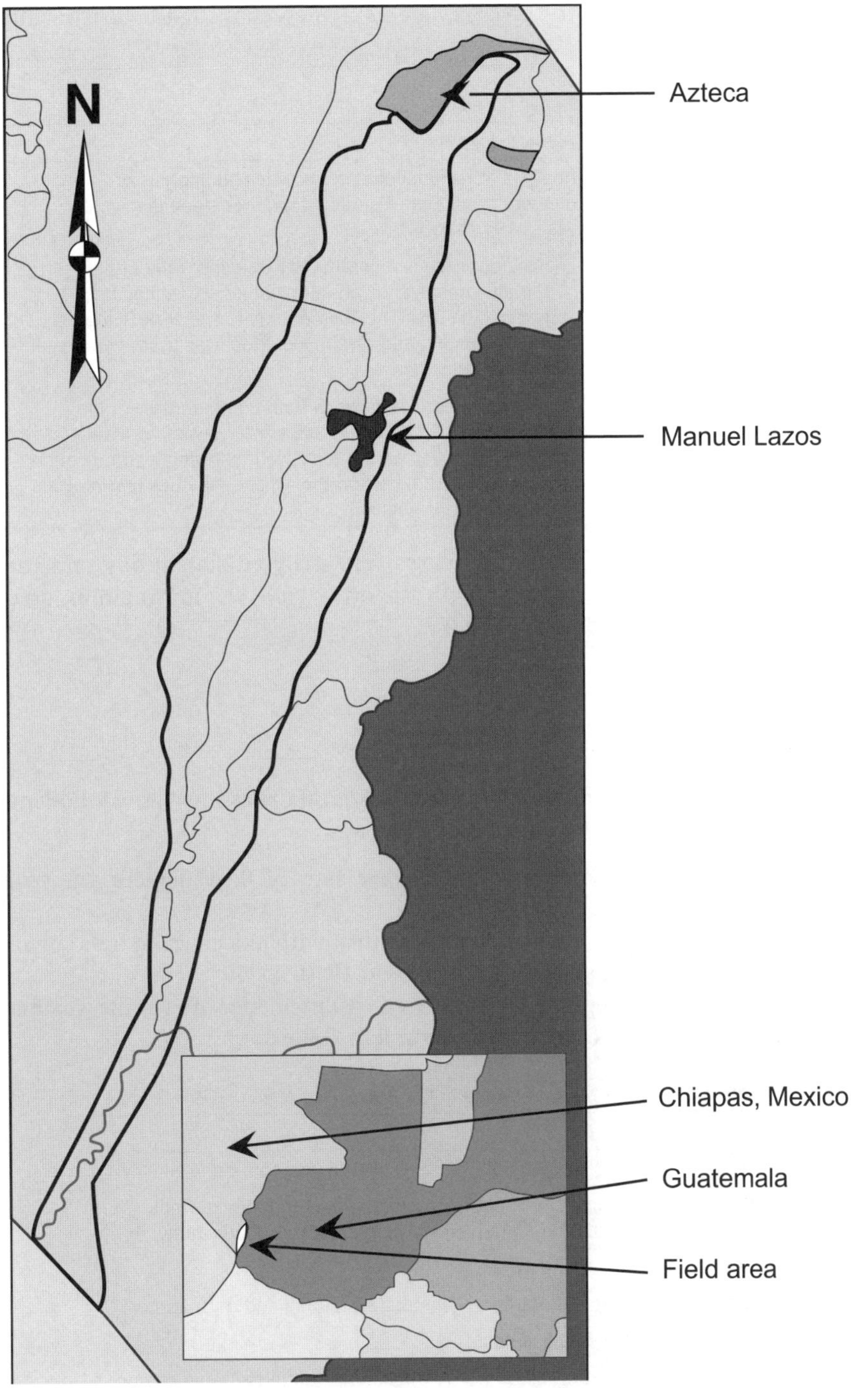

Figure 5.2 The location of the field validation work

The different steps of the process were implemented using the instruments summarized in Table 5.1.

Table 5.1 Summary of methodologies used

Stage	*Methodology*
Exposure	Analysis of regional climate models and analysis of meteorological data from local weather stations (over 40 years, 1970–2009).
Sensitivity and perception	Focus groups in each community to identify the effects of climate change and variability on local livelihoods and activities. The CRiSTAL (Community-based Risk Screening Tool – Adaptation and Livelihoods) tool was used, combined with other questions.
Perception and adaptive capacity (all steps)	Semi-structured interviews in both communities. In the Azteca *ejido*, all 46 families were interviewed; in Manuel Lazos, 51 families were interviewed following a stratification of the population based on their main activities (*estrategias de vida* or life strategies).

The information from the interviews was analysed statistically and the outputs of this analysis, together with the other collected information, used to generate the results described below.

Results

Exposure

The analysis of climate models and meteorological data showed the following situation and trends for this region of Chiapas:

- Temperature is increasing at an average rate of 0.01 degrees per year (1970–2009).
- Rainfall has a strong inter-annual variability, with a decreasing trend of an average of 11.5 mm per year in the period 1970–2009.
- There is a local perception about the frequent incidence of extreme weather events in the last decade, as summarized in Table 5.2.

Table 5.2 Extreme weather events in the Soconusco region, 1998–2010

Year	*Event*
1998	Extreme drought. The rainy season was delayed until July.
2005	Tropical storm Stan. Extreme rainfall and strong winds. Worst flooding in memory in the lowlands (Tapachula city).
2006	Very strong winds.
2007	Hurricane Barbara. Very intense rainfall and dry summer.
2008	Very strong winds.
2010	Late rainy season. Very intense rainfall in the rainy season.

Source: Prado, 2011

Sensitivity

The first task was the identification of the key life strategies (*estrategias de vida*) and their components.

In *ejido* Azteca, the key livelihood components for income were coffee, payments for environmental services (the *ejido* maintains a protected forest) and governmental plans oriented to different groups (such as young mothers, children, one-parent families, etc.). These three activities were common to all families. Each family also had one additional activity: maize cultivation, remittances from abroad or daily farm work on other farms.

The situation in *ejido* Manuel Lazos was different because of its proximity to a town. The basic common livelihood components here were coffee and/or rambutan (a tree fruit) combined with government plans. All families had in addition one or more of the following: remittances from abroad, daily work on other farms and town jobs (governmental, private, domestic and other).

Table 5.3 shows the results of the focus groups and the application of CRiSTAL, indicating climate effects over key productive activities for local livelihoods.

Table 5.3 Climate effects on the main productive activities and resources

Activities/resources	*Climate effects*	*Importance*
Crops (coffee, rambutan, maize)	High incidence of diseases in wet years Loss of flowers Low yields or total crop loss	Very important
Access road	Damage and/or total loss of access	Very important in Azteca
Soils	Strong soil loss in heavy rains	Very important
Infrastructure for coffee drying	Damage	Important
Rambutan and coffee prices	Instability	Important
Governmental programmes	Increase in bad years	Important

Source: Prado, 2011

Adaptive capacity

The results from processing the information gathered in the interviews in relation to the different steps proposed in the analytical framework are as follows.

Perception. The results from interviews about perception of the effects of climate change and variability show that the local population is reasonably aware of them. Table 5.4 shows the percentage of the interviewed families that perceive different effects on different aspects of their lives.

Table 5.4 Local perceptions of climate effects on the main crops and infrastructure

Ejido	*Event*	*Maize*	*Rambutan*	*Coffee*	*Houses*	*Community infrastructure*	*Soils*
Azteca	Intense rainfall	Plant disease (48%) Plants topple (20%)		Plant disease (63%) Fruits fall (15%)	Houses affected by landslides	Roads damaged every year	Soil loss (100%)
	Wind	Harvest loss (61%)		Fruits fall (69%)			
	Tropical storm Stan	Harvest loss (73%)		Harvest loss (100%)	Houses affected (roofs, flooding) (30.4%)	Houses damaged (30%) Roads damaged	Soil loss (52%)
Manuel Lazos	Intense rainfall		Flowers fall (70%)	Plant disease (12%) Fruits fall (67%)	Flooding	Secondary roads damaged	
	Drought		Harvest loss (30%)	Harvest loss (69%)			
	Tropical storm Stan				Houses damaged (10%)	Road bridge destroyed	Soil loss (24%)

Note: Perceptions without an associated percentage were perceived by less than 10% of families.

Source: Prado, 2011

Reaction. The analysis of this factor provided some unexpected results. In both communities there was a high percentage of respondents (much higher in the more isolated community) who thought that there is nothing they can do about the effects of climate change. Several reasons were given to justify their answers, ranging from 'who are we to change God's will?' (or nature) to the need for high levels of investment or the lack of resources. Since in one of the cases (Azteca, where 100 per cent of the families were interviewed) almost half of the families had this attitude, it is clear that any intervention aimed at climate change adaptation in this area needs to pay attention to this issue.

Table 5.5 Local reactions to the possibility of adaptation

	Azteca (%)	Manuel Lazos (%)
Yes, it is possible to take action to reduce the impacts of climate change	54.0	71.0
No, actions cannot be taken to reduce the effects of climate change	46.0	29.0
Justification of negative answers (% related to negative answers):		
• God's will	38.1	35.7
• Nature	61.9	35.7
• High level of investment required	–	21.4
• Lack of resources	–	7.1

Source: Prado, 2011

Identification of actions. Those families that believed it is necessary to take action to reduce the effects of climate change and its variability on their livelihoods put forward a relatively broad range of possible actions to do so. At the same time, it seems that these actions were not very consistent with the type of effects they identified during the analysis of their perceptions of the effects of climate change.

Table 5.6 shows the identified actions and the percentage of families mentioning them in the interviews.

Table 5.6 Potential adaptation actions identified by local communities

Actions	*Azteca (%)*	*Manuel Lazos (%)*
Reforestation	55.0	13.9
Soil conservation practices	40.0	19.4
Irrigation	–	30.6
Emigration	4.6	–
Soil retention walls	4.0	11.9
Crop fertilization	–	8.5
Crop diversification	–	5.6
Community organization	–	5.6

Source: Prado, 2011

While the families identified valid adaptation actions, it seems that their range is narrow, particularly in Azteca, where they appeared to be limited to just soil conservation. The situation in Manuel Lazos was better but still narrow compared with the type of climate effects perceived.

Preparedness to implement actions. In this step, the application of the framework faced some difficulties, basically because the proposed concept did not adequately address the issue of decision making. The initial concept was to analyse the organizational level, the planning capacity and the capacity for lobbying and getting financial resources and technical assistance.

In the particular case of these communities, some other elements arose that had to be considered, such as the level of action by governmental organizations (which are very active and have significant financial resources in the case of Mexico) and, related to this point, who decides what in this context.

Local organization. Based on the description of the results of the planned steps, it was found that local organization was very good in both communities due to the existence of the *ejido* organization and other local groups, as shown in Table 5.7.

Table 5.7 Local organizations active in the communities

Type of local organization	*Azteca*	*Manuel Lazos*
Assembly of the *ejido* (monthly meetings of all members)	X	X
Group of women beneficiaries from governmental programmes	X	X
Forest group beneficiaries from the PES[1] programme	X	–
School committee	X	X
Water committee	X	X
DICONSA[2] rural store committee	X	X
GRAPOS[3] (NGO of organic producers from several communities in the region)	X	X
Cacao and rambutan producers' association (farmers from different communities)	–	X
Union of *ejidos*	–	X

1 Payment for Environmental Services, a programme of governmental subsidies managed by CONAFOR (Comisión Nacional Forestal/National Forestry Commission).

2 Distribuidora Conasupo SA, a governmental chain of food stores across Mexico to ensure food supply to rural communities. CONASUPO was the Compañía Nacional de Subsistencias Populares (National Company of Popular Subsistence), a governmental organization distributing agricultural subsidies until it was converted into DICONSA a few years ago.

3 Grupo de Asesores de Producción Orgánica y Sostenible (Advisers Group for Organic and Sustainable Production).

Source: Prado, 2011

Planning capacities. Local planning capacities range from very low to non-existent; there were several projects in these communities but they were planned by organizations and NGOs with or without the communities' participation. Even when there was participation, the local groups were not trained in basic planning tools. Moreover, when they had initiatives of their own, they needed to find an

NGO or governmental organization willing to help with the planning. There are consultants in the cities (Tapachula and others) but the community groups seemed not to have the financial resources or the clarity of thinking to hire them.

Decision making. The situation described above, together with the overall context noted here, made it necessary to explore an issue that was missing from the initial approach: who takes the initiative and decides about projects and activities in these communities – the communities, through their organizations, or the external stakeholders (government agencies, NGOs, projects, etc.) active in the region? The importance of understanding this issue becomes relevant later when analysing the implemented actions and their sustainability.

The exploration of this issue through the interviews led to the results shown in Table 5.8.

Table 5.8 Source of decisions about projects and activities

	Azteca (%)	*Manuel Lazos (%)*
Initiatives originated and decided by the community organizations	13	29
Initiatives originated and decided by external organizations	87	70

Source: Prado, 2011

This issue became critical in terms of understanding the relative importance of the different constraints, and is one that merits further investigation. While these results showed the significant prevalence of external initiatives, the proportion of local initiatives that were implemented by external sources of technical and financial support remains to be explored. In other words, was the small number of initiatives based on community decisions implemented because external sources of support respected them, or simply because they coincided with their priorities?

The evidence on this issue is very important in terms of assessing the constraints of local adaptive capacity to climate change and variability; it will indicate whether the initiative for the adaptation process lies in the hands of the communities, or in the hands of the external sources of funding and technical expertise.

Effective implementation and sustainability of actions. There has been no action taken yet in these communities on climate change adaptation, and so an analysis of the results of another governmental programme on soil conservation practices was carried out, as these practices are also representative of some of the actions required for climate change adaptation in these hilly and rainy areas. The soil conservation practices implemented in both communities included terraces, hedgerows, retaining walls with live or dead materials, and contour ditches.

These practices were perceived as successful by the farmers, as shown in Table 5.9, and some farmers identified more than one benefit from them.

Table 5.9 Benefits generated by soil conservation practices

Benefit	*Farmers perceiving the benefit (%)*	
	Azteca	*Manuel Lazos*
Litter retention (dead leaves, small branches, etc.)	70	58
Soil retention	32	16
More vigorous crops	27	25
Higher crop yields	12	0
Improved soil moisture	5	21

Source: Prado, 2011

Unfortunately, the maintenance of these conservation practices and structures was not as good as expected, reinforcing the importance of assessing sustainability. In the case of Azteca, the conservation of terraces and walls was implemented by 76 per cent of the farmers, a figure that drops to 46 per cent in Manuel Lazos. Moreover, in both cases the maintenance of contour ditches ceased completely as soon as they became silted up with sediments.

The reasons for the decline in maintenance are shown in Table 5.10, reinforcing the need to explore thoroughly issues regarding decisions and commitment.

Table 5.10 Reasons for the decline in maintenance of soil conservation practices

Reasons for decline in maintenance	*Azteca*	*Manuel Lazos*
Lack of governmental support	38	16
Decomposition of materials (wood mostly)	38	46
Lack of time	24	38

Source: Prado, 2011

The reason 'lack of governmental support' clearly points to a perception that these actions benefit government, not the local communities. The other two reasons can also be related to the feeling that the conservation practices were of low priority to the communities, perhaps because they were not their own initiatives, or perhaps because of other causes that need to be identified to clarify these key issues.

The impact of climate change and variability effects. As in the previous step, it was not possible to assess this aspect as there were no climate change adaptation activities. The process with the soil conservation practices described in the previous section shows that the farmers perceived the benefits and were conscious of the changes, but that unfortunately this perception was not enough for many of them to maintain the improvements despite the benefits.

Again, the possible causes of this apparently contradictory behaviour were not clear; the main hypothesis of the authors relates to issues of ownership of the initiatives and who are the decision makers; in turn, these issues may have their roots in the lack of local planning and lobbying capacities and/or the level of attention paid by external sources of technical assistance and funding to initiatives that are not their own. This study did not throw much light on these hypotheses, and therefore they need to remain hypotheses until further studies prove or reject them.

Disengagement of the population along the process. For some communities the population was disaggregated to show existing bottlenecks and bifurcations along the route. In the case of these communities, this analysis led to the results in Table 5.11.

The table shows the percentage of families entering one particular stage, the proportion of them that were not able to go beyond that stage (retained at or blocked by the characteristic issues of that stage), and the proportion that continued to the next stage.

Table 5.11 Gradual disengagement of people during the process

	Azteca			*Manuel Lazos*		
Stage	*Families beginning the stage (%)*	*Families blocked at this stage (%)*	*Families ending the stage (%)*	*Families beginning the stage (%)*	*Families blocked at this stage (%)*	*Families ending the stage (%)*
Perception	100	20	80	100	30	70
Reaction	80	36	44	70	21	49
Actions – identification	44	22	22	49	14	35
Actions – implementation	22	17	5	35	10	25

The percentages in the table are directly based on the collected information in some cases and on the author's estimation in others. At the stage of perception, the proportion of families blocked was an estimation based on the level of perception of the effects of climate change. These perceptions were more widespread in Azteca than in Manuel Lazos, leading to a higher level of retention in the latter and signalling the need for action in this area in both communities, but with a greater emphasis on Manuel Lazos.

At the stage of reaction, the level of retention was taken directly from the collected and processed information presented in Table 5.5. The proportion of retained families at the stage of identification of actions was estimated based on the information summarized in Table 5.6 and in the subsequent comments presented in the text. At the stage of implementation of actions, the percentages were taken directly from the information collected, processed and presented in Table 5.10.

The main purpose of this analysis is not to get highly accurate retention figures at each stage but rather it is about fostering a reflection about the proportion of the local population that has capacity to reach a meaningful implementation of actions (as opposed to merely doing what they are told, or are paid to do). It is expected that this type of analysis will help organizations and projects achieve better targeting of their activities and audiences, as can easily be deducted from an analysis of the differences between these two communities.

Analysis of the adaptive capacity of the communities

Based on the collected information and analysis, it is possible to conclude that the adaptive capacity in these communities is low. There is still room to improve the perception of the problems, the passive reaction to the problems is a matter of concern, there are some basic ideas about what to do but these are fairly inadequate, and the level of organization is good but the planning and lobbying capacity is low. Moreover, it is clear that most of the initiatives in these communities have an external origin and there are some questions not sufficiently well answered about the level of maintenance and sustainability of past interventions whose benefits were clearly perceived as positive by these communities.

While this overall assessment is valid for both communities, it is also evident that Azteca is more vulnerable than Manuel Lazos. While both have similar exposure, the sensitivity of Azteca is greater because it is more dependent on agricultural activities and its topography is steeper. Moreover, in almost all aspects analysed for adaptive capacity, Azteca is performing at a lower level than Manuel Lazos. Therefore, in terms of prioritizing intervention in the most vulnerable communities, the efforts should be focused on Azteca.

Use of the results to design interventions

The results generated along the process should be used to understand what the situation is and also to inform decisions about the different types of intervention required.

One of the interesting features of the proposed process is that the systematic identification of different problems and constraints leads to the identification of specific actions aimed at each problem and constraint. Table 5.12 summarizes some of the actions to be considered in the case of the studied communities.

Table 5.12 Potential actions to foster adaptive capacity

Problem/constraint	*Possible actions*
The effects of climate change are not perceived	Awareness development, environmental education, sensitization, etc.
Passive reaction (the effects are perceived but it is not believed that adaptation actions are feasible)	Visits to other areas to see and understand how other communities are reacting to climate change effects, sensitization of religious leaders, presentation of government programmes, etc.
Lack of clear ideas and proposals about what to do to adapt	Training, visits, exchanges, etc.
Lack of organization and/or weak organizations	Strengthening local organizations, particularly in aspects such as planning, lobbying, funding and technical assistance negotiation, and other issues.
Lack of capacity or weak capacity for lobbying	Access to decision makers, contacts, exchanges with successful organizations, negotiation of sponsorships and twinning arrangements with other, stronger organizations for training and related issues, etc.
Ineffective implementation and monitoring	Technical assistance, coaching and training on project implementation, monitoring, self-evaluation and adaptive management.

Conclusions and lessons learned

The results of the field testing show very clearly the different nature of the constraints and allow for the identification of specific actions to be taken to address these barriers specifically. The barriers ranged from lack of perception to beliefs about the uselessness of acting against God or nature, and to a lack of capacity and political clout, among other things.

Therefore, it is concluded that this first validation in two communities showed that the proposed framework is a useful and promising starting point to understand adaptive capacity in a way that allows for actions directed at specific constraints.

Key missing aspects were the issue of ownership and decision making, as well as issues of identification, implementation and sustainability of the adaptive actions described. Therefore, this is an issue to be considered in further developments of the approach.

It can also be argued, and the authors agree with this as a preliminary judgement, that the procedure is too laborious, particularly because of the high number of family interviews that were conducted. This is a procedural issue and not a conceptual one, and therefore it does not invalidate the framework, but additional work is definitely required to find adequate ways of completing this step in a more efficient and less time-consuming way. Improvements to the process are currently being developed and are expected to be field tested in the near future.

References

CRiSTAL (2012) CRiSTAL (Community-based Risk Screening Tool – Adaptation and Livelihoods). Available from <www.iisd.org/cristaltool/download.aspx> [accessed 28 April 2012].

Flora, C.B., Flora, J.L. and Fey, S. (2004) *Rural Communities: Legacy and Change*, 2nd edition, Boulder, CO: Westview Press.

Imbach, A.C. (2012) *Estrategias de vida: Analizando las conexiones entre la satisfacción de las necesidades humanas fundamentales y los recursos de las comunidades rurales [Life Strategies: Analysing the Links Between Fundamental Human Needs and the Resources of the Rural Communities]*, Turrialba, Costa Rica: Geolatina Ediciones.

Prado, P.F. (2011) Diseño e implementación de una metodología participativa de diagnóstico de la capacidad adaptativa a la variabilidad climática en la cuenca del Cahoacán, México *[Design and Implementation of Participatory Method to Assess Adaptive Capacity to Climate Change in the Cahoacan Watershed, Mexico]*, MSc thesis, Centro Agronómico Tropical de Investigación y Enseñanza (CATIE).

About the authors

Alejandro C. Imbach was born in Argentina (in 1949) and has lived in Costa Rica since 1985. He is an agronomist with an MSc in natural resources management. After working with small farmer co-operatives in northern Argentina, he was Professor and Dean of the Forestry College in the University of Misiones, Argentina. He joined the International Union for Conservation of Nature (IUCN) Central America in 1987 and worked for IUCN in various capacities. In late 2004 he set up his own consultancy in planning, monitoring and evaluation. In 2009 he joined the CATIE graduate school, launching a master's degree programme in development practice and lecturing on topics related to rural development and planning.

Priscila F. Prado Beltrán is an agronomist with an MSc in ecological agriculture. She joined the Community Network of Natural Resources Management (MACRENA) to work with farmers and local communities in the northern Andean areas of Ecuador, providing technical assistance, organizing field schools and training trainers in maize agro-ecological management. After completing her master's degree, she has been working as a consultant developing land use plans for different municipalities in Ecuador.

CHAPTER 6

The role of policies and institutions in adaptation planning: experiences from the Hindu Kush Himalaya

Neera Shrestha Pradhan, Vijay Khadgi and Nanki Kaur

The study was conducted in the Hindu Kush Himalaya to analyse the role of policies and institutions in adaptation planning. The study focuses on four entry points identified as relevant to local adaptation: agroforestry, water governance, flood mitigation infrastructure and migration. This paper attempts to analyse the findings of case studies of community-based adaptation conducted by the International Centre for Integrated Mountain Development (ICIMOD), the International Institute for Environment and Development (IIED) and the Stockholm Environment Institute (SEI) together with ICIMOD's regional partners in China, India, Nepal and Pakistan to determine the policy interfaces between public and non-public institutions in order to support (or not support) local adaptation strategies. The evidence suggests that policies and institutions are not always supportive of local practices – and drivers other than climatic variability actively influence changes (increasing or decreasing people's vulnerabilities to climate stress) at the local level that are not always climate proof. The paper validates and strengthens the needs for well researched government decisions that take account of local practices and consider various drivers of change.

Keywords: institutions, adaptive capacity, adaptation planning, policy, markets, Hindu Kush Himalaya

The Hindu Kush Himalaya (HKH) region, also known as the Third Pole, is the source of 10 major river systems in the region and provides water for irrigation, power and drinking for 1.3 billion people – over 20 per cent of the world's population (Pradhan et al., 2012). Extending from Afghanistan in the west to Myanmar in the east, HKH covers eight countries, and the region's fast-growing and rapidly urbanizing population is creating pressure on water and other environmental resources.

The Fourth Assessment Report of the Intergovernmental Panel on Climate Change (IPCC, 2007) reported that global warming and climate change are impacting on mountains (Bandyopadhyay and Gyawali, 1994; Hua, 2009; Scheraga and Grambsch, 1998; Xu et al., 2009) and mountain ecosystems (ICIMOD, 2010). Significant warming is predicted in south Asia and on the Tibetan plateau. The resulting accelerated rate of glacial retreat will alter

http://dx.doi.org/10.3362/9781780447902.006

the contribution of glacier melt water, affecting high altitude wetlands and changing the base flow of HKH rivers. The next three to four decades will see an increase in base flow, and then a decrease, with effects on agriculture (Rasul, 2010), ecosystems and their services (Malone, 2010), food security (Nellemann and Kaltenborn, 2009) and energy security (Asia Society, 2009). This cascade of impacts will affect the livelihoods of people living in the mountains and those downstream (Xu et al., 2009), especially women.

Water-induced hazards and stresses are already an increasing threat in the HKH region. In some areas, water is chronically scarce; in others, people have lived with recurrent floods and droughts for centuries. Local economies are largely agriculture-based and highly dependent on natural resources such as water, soil and forests. With increased demand and competition for water, and more variability in its availability, people are struggling to manage traditional arrangements for dealing with the water scarcity with which they have always lived. At the sub-national or local level, extreme and non-extreme weather or climate events increase the vulnerability to future events by affecting livelihood options and resources, and by minimizing the capacity of societies and communities to prepare for and respond to future disasters (IPCC, 2012: 1–19). Households and communities in the region have evolved their own strategies to cope with, and adapt to, periods of flooding and drought, thereby enhancing their resilience (Pradhan et al., 2012).

At scales ranging from regional to local, diverse sets of actors in the countries of the HKH region are beginning to develop plans for dealing with the effects of climate change on the economy and on livelihoods. Such plans need to be informed by an empirical understanding of how effective adaptation can take place. The first round of research contributed to that understanding by demonstrating that local institutions play a critical role in enhancing people's capacity to respond to too much and too little water at household and community levels. The findings also pointed to the influence of national policies and institutions on the development of adaptive capacity, and to the frequent disconnects between national and local plans and institutional options for strengthening adaptive capacity (ICIMOD, 2009). The second round explored these local–national interactions in more depth, to understand better how effective interactions between local and national institutional contexts can be fostered for building adaptive capacity.

This chapter outlines the findings of this research conducted by the International Centre for Integrated Mountain Development (ICIMOD) together with the International Institute for Environment and Development (IIED), the Stockholm Environment Institute (SEI) and national partners in China, India, Nepal and Pakistan to demonstrate the extent to which local households and communities are spontaneously adapting, and will continue to adapt, through actions that are independent of planned programmes and policies. These actions are sometimes referred to as 'autonomous', as opposed to planned, responses to climate change and are influenced by external environment factors such as policies and institutions.

Approach and methodology

Adaptation to climate change is highly local and its effectiveness depends on local institutions through which incentives for individual and collective actions are structured. It is important to understand the role of institutions, if adaptation to climate change is to support vulnerable social groups at the local level (Agrawal, 2010). While considering that the role of institutions is a specific component of adaptive capacity and acts as the means of delivery of external resources to facilitate adaptation (Agrawal, 2010; Dovers and Hezri, 2010; Christoplos et al., 2010), this study analyses the role of policies and institutions in local adaptation planning.

The study was conducted in collaboration with international resource centres and the ICIMOD regional member country partners in their respective countries.

Study area

The study was conducted in various communities in four different countries in the HKH region (Figure 6.1), with a focus on four thematic issues as key entry points: agroforestry, water governance, flood mitigation infrastructure and migration. A comparative study was conducted in Baoshan municipality

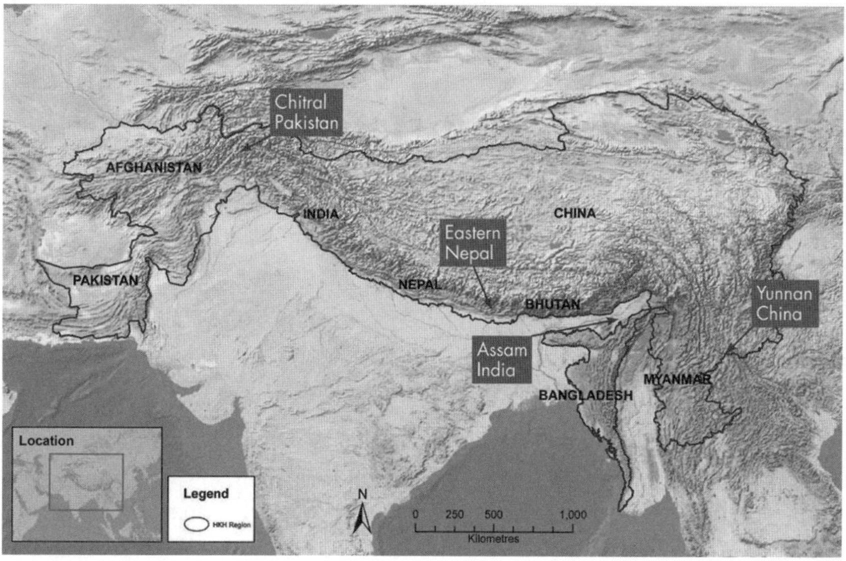

Figure 6.1 Sites selected for the case studies

Source: ESRI

in Yunnan (China), Mustang district (Nepal), and Khyber Pakhtunkhwa province and Azad Jammu and Kashmir (Pakistan) to understand the role of agroforestry diversification and intensification and the institutions involved in supporting the adaptive capacity of communities.[1] In Mulkhow and Shishikoh valleys, Chitral district (Pakistan), small-scale water management and the role of local institutions were studied to analyse the role of community-based and government organizations in adaptation planning.[2] Lakhimpur and Dhemaji districts, Assam (India) and Koshi (Nepal) were studied to analyse the role of infrastructure measures, including embankments to adapt to floods.[3] Chitral district (Pakistan), Assam (India), Dhankuta, Sunsari and Saptari districts (Nepal) and Yunnan (China) were considered in order to study the role of migration and the links between water stress and remittance as a measure of livelihood diversification to adapt to climate change and other drivers.[4]

A common methodological and conceptual framework was prepared. The baseline data were collected using secondary resources such as literature review, scientific journals, published and government reports, and climate and hydrological data. A purposive sampling strategy was used to select representative communities in the countries based on the previous study conducted to scope the major issues and community-based adaptation strategies (ICIMOD, 2009). Some of the major issues identified by the study in the HKH region were policies and market change, local politics and governance, livelihood diversification, the role of social organization, social capital and networks and gender considerations (especially impacts on women). Participatory rural appraisal (PRA) and rapid rural appraisal (RRA) techniques were followed in addition to focus group discussions, observations and case studies to collect data. The analysis of the collected data was conducted based on the conceptual framework to evaluate the role of policies and institutions in adaptation planning (Figure 6.2).

Conceptual framework for adaptation planning

The role of institutions, as a specific component of adaptive capacity and in mediating adaptive capacity, has been recognized earlier in this study and in broader literature (Agrawal, 2010; Christoplos et al., 2009; Dovers and Hezri, 2010). Institutions are defined here as 'commonly understood rules and norms that stipulate what actions are required, permitted, or forbidden in a particular situation' (Poteete and Ostrom, 2004). They may be informal (for example, social norms and taboos) or formal (for example, constitutions and property rights regimes) and exist at multiple scales of human organization. The manifestation of institutions into specific forms, such as departments, associations or agencies, refers to the *organizational* dimension of institutions. These could include public (state/government), private (market/service) and civic (civil society) organizations.

There is a lack of consistency in the way in which institutions have been defined in the context of adaptation; a lack of clarity around their role in

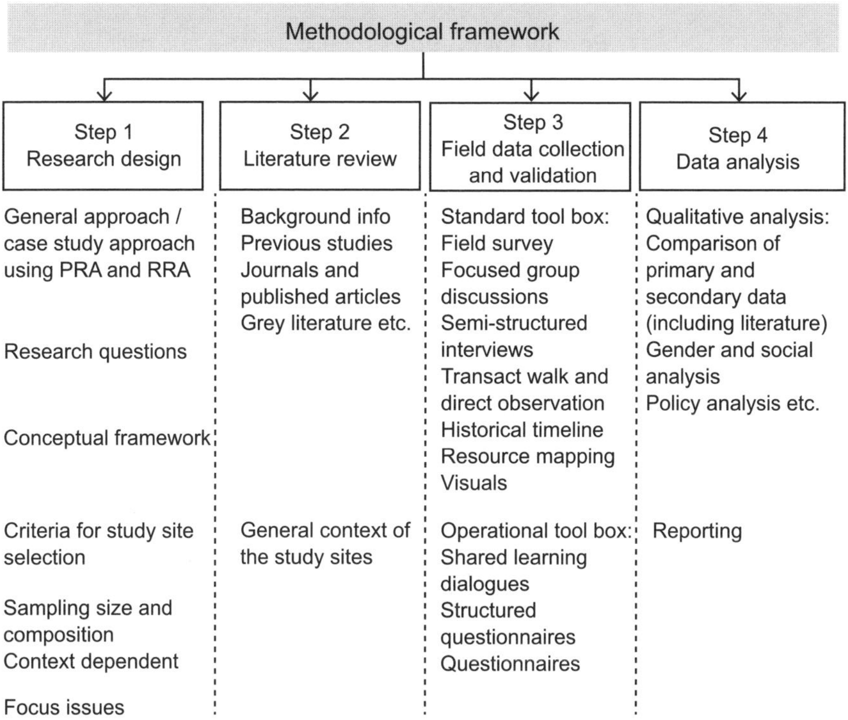

Figure 6.2 The methodological framework

facilitating adaptive capacity; and a lack of focus on how to move from a discussion of what should happen – adaptation outcomes – to a discussion about how this should happen, or what institutional design is required for delivering effective adaptation outcomes (Dovers and Hezri, 2010).

Adaptation planning aims to enhance the effectiveness of institutional systems in building adaptive capacity. This requires clarity around the specific roles of different institutions and the ways in which they interact with one another to enable or disable the development of adaptive capacity. The conceptual and analytical framework presented in this paper aims to address this gap in understanding, particularly in terms of the interfaces between formal and informal institutions, and local and national scales. The conceptual framework starts with the hypothesis that adaptation effectiveness is determined by the interface between formal and informal, and between public, private and civic institutions that operate at different scales of adaptation planning (see Figure 6.3).

Different types of institutions have specific comparative advantages in terms of the delivery of aspects of adaptive capacity. Agrawal's (2010) analysis of the United Nations Framework Convention on Climate Change (UNFCCC) database on adaptation sets out the comparative advantages of some public, private and civic institutions in the delivery of adaptive capacity (Table 6.1).

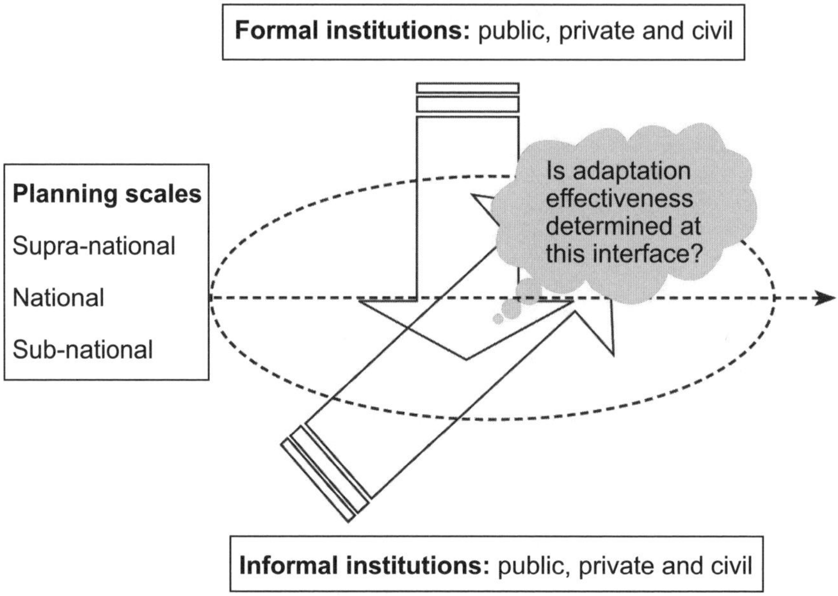

Figure 6.3 The conceptual framework
Source: Pradhan et al., 2012

Table 6.1 Comparative advantage of types of institutions in building adaptive capacity

Institutional realm	*Comparative advantage vis-à-vis adaptive capacity*
Public e.g. bureaucratic agencies and elected bodies	Public sector institutions are more likely to facilitate adaptation strategies relating to communal pooling, diversification and storage owing to their command over authoritative action, and their ability to channel technical and financial inputs into rural areas
Private e.g. service organizations and private businesses	Private sector organizations, because of their access to financial resources, are likely to have greater expertise in promoting market exchange and diversification, but may also be able to advance communal pooling if one takes into account not-for-profit service organizations
Civic e.g. non-governmental, co-operative and membership organizations	Civic sector institutions can strengthen different adaptation responses due to their greater flexibility in redefining goals and adopting new procedures

Source: Agrawal, 2010

Agrawal's analysis also suggests that civic and public institutions are the ones most commonly involved in facilitating adaptation to climate change, while private sector and market institutions have played a relatively small role in facilitating or reinforcing adaptation. Furthermore, at the local level, civic institutions, when functioning on their own, tend to be informal institutions.

However, when public institutions are involved in adaptation they tend to collaborate with formal civic institutions rather than strengthening informal civic institutions. In adaptation planning, the strong partnership and collaborative arrangements between or among the private, public and civic institutions (Figure 6.4) are extremely important in addressing climate hazards and related adaptation (Agrawal, 2010).

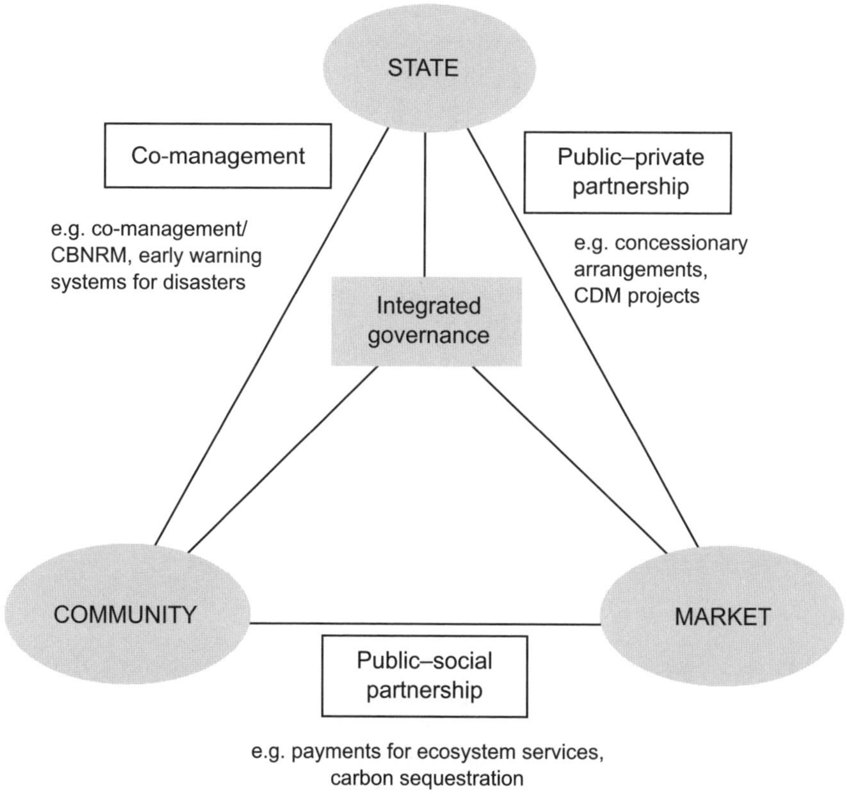

Figure 6.4 Schema of collaborative institutional arrangements
Source: Agrawal, 2006

Depending on the extent to which there is a match between the aims and comparative advantages of different institutions, the interface between institutions can be supportive, non-supportive or non-existent (Figure 6.5). An aim of adaptation planning should be to enable and facilitate supportive interfaces where possible.

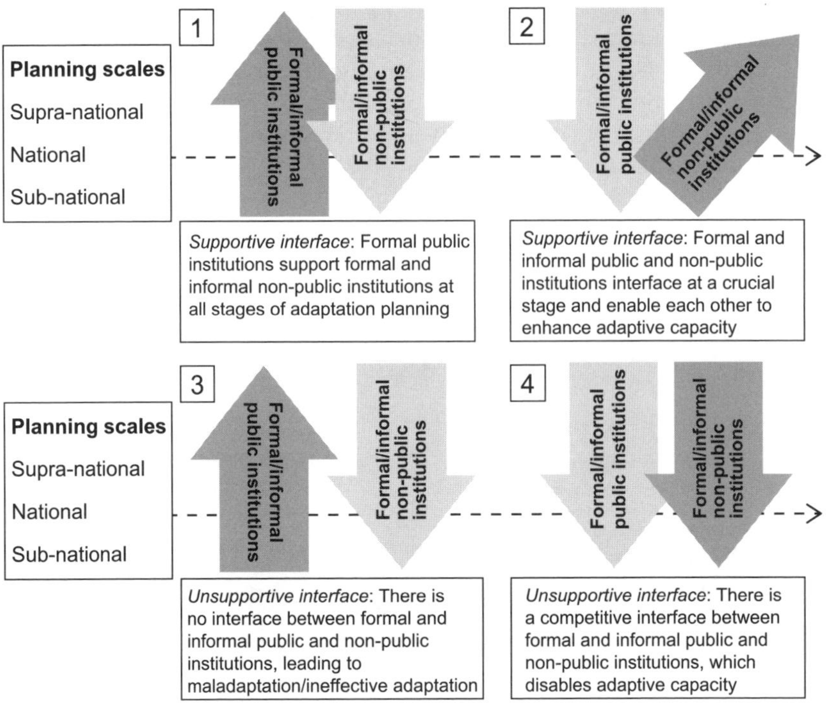

Figure 6.5 Examples of adaptation planning interfaces

Source: concepts developed by Kaur and Goeghegan (2010) as cited in Pradhan et al., 2012

Analysis of adaptation planning interfaces: case studies

Analysis of the role of institutions

Interfaces between public and non-public institutions are formed through government policies and programmes that are implemented at local levels. Frequently, policies are designed to support CBA, and in theory should complement local needs and adaptation strategies. For example, national policies on water management in Pakistan are premised on the participation of communities at all levels of management: the conservation of water resources; increasing the coverage of water supply and treatment facilities; the decentralization of planning, development and management to the local level; and the inclusion of all stakeholders (Pradhan et al., 2012). In addition, flood management policies in Assam in India and a tree crop programme in Baoshan in China have been developed with the objective of enhancing community capacity and resilience to cope with climate variations that result in floods and droughts.

A more realistic interface is where public institutions interface with non-public institutions at crucial stages. Such interface is generally responsive to emerging changes, and targeted to improve the communities' adaptive capacities. The Baoshan municipal government in China, in an effort to curb soil erosion, initiated the expansion of walnut production with the goal of planting over 30,000 hectares of walnut trees in Langyang district alone. The government also provided technical support for planting, with specific guidelines on the suitability of different walnut species at different elevations. Likewise in Nepal, farmers have been provided with seeds and market access to shift apple production to higher altitudes in order to adapt to increasing temperatures. Although the government programmes and policies are supportive in enhancing the adaptive capacity of local communities, they are more driven by economic benefits and are not responsive to agricultural diversification strategies for adaptation. State and market institutions work in tandem, but markets respond more quickly to crop successes and failures that may be due to shifts in climate.

The government-led water management system that has replaced traditional community-based water management systems in Chitral, Pakistan was implemented without much community participation. It introduced different water prices based on the type of crop – water for food crops is cheaper than water for other crops – in order to support government policy on food security. Although the Department of Irrigation constructed 15 irrigation channels, their maintenance was handed over to contractors who were not monitored. The system lacked transparency and accountability, resulting in limited community participation and limited support for CBA needs. As a result, the communities took up the ongoing work of maintaining the system in order to avoid the risk of crop failure. The unsuitability of policies to local conditions often affects their effective implementation. Although policies and programmes may exist, they may not always interface with non-public institutions if the interfaces are not formalized and supported by implementation guidelines. Therefore, unclear roles and lack of understanding between formal and informal institutions can result in maladaptation.

A supportive interface where formal public institutions support formal and informal institutions at all stages of adaptation planning is most desirable, but rare to find in reality. In Mulkhow, Chitral in Pakistan, water distribution is managed by communities, a system established by its former rulers. The role of the state terminated after handing over the water distribution rights to communities, usually a cluster of villages, represented by informal institutions. Village representatives agree on water rights between villages and, at the village level, representatives of each tribe or clan enforce the decisions based on individual water rights. AKRSP, a non-public formal institution, helped formalize and modernize the system by making it equitable in terms of allocating water according to the contribution of each household towards renovation and maintenance of the system. This system proved to have a supportive interface between the public and non-public institutions, confirming that the success of historically developed adaptation practices depends crucially on the nature of prevailing formal and informal rural institutions (Agrawal et al., 2008).

Many instances of policies or programmes can be found that are designed without any consideration of local needs or practices. For instance, the government-based water distribution programme in Chitral in Pakistan introduced a more scientific water distribution system by disregarding entirely and replacing the effective traditional systems without community participation.

In Assam, India, flood mitigation infrastructure is entirely led by formal public sector state institutions without much scope for local participation or the inputs of informal non-public institutions. Although the first flood management policy in 1953 recommended the construction of embankments as only a short- and medium-term plan, embankments were promoted widely as they were relatively cheap and provided more visibility than the recommended long-term measures such as storage dams. Although initially effective in protecting people from floods and providing alternative livelihood options, the embankments deteriorated due to inadequate maintenance over several decades. The increased dependence on embankments for flood protection increased the vulnerability of communities. This demonstrates an unsupportive interface between the formal public sector and informal non-public sector, which exist in completely separate domains within the same geographic area.

The study in the HKH region showed that migration is considered as one of the strategies to adapt to socioeconomic and climatic changes. Over the last decade, the concept of adaptation has gained prominence in the climate change discourse, particularly within the UNFCCC agenda and the Copenhagen Accord. The countries in the HKH region have different policy approaches for labour migration at the national level, but they usually support international labour migration. However, internal migration has gained less recognition at the formal institutional level. At the community level, people are more attracted to migration as it diversifies their income and brings greater economic benefits to their family to cope with changes. Besides remittances, they also bring back skills, which enhance their adaptive capacity. A supportive interface exists in Nepal, where migrants who work abroad support the national economy, contributing 22.9 per cent of gross domestic product in 2010. While national legislation provides a framework for the protection of this group of labour migrants, there is no guarantee that the authorities concerned will abide by all the legislative provisions. In this case, there exists an unsupportive interface between formal public and informal non-public institutions, which undermines the role of extended family networks, kin and clan networks, hometown associations, and other such social networks in providing social protection to the migrant and the non-migrating members of a family.

Analysis of policy implementation

Recent research suggests that existing institutions are unlikely to be able to cope efficiently and equitably with climate change, particularly in developing countries (Kane and Yohe, 2000; Kates, 2000 as cited in Tompkins and Adger,

2003), partly because climate policy decisions are mostly made at the national level, although the consequences of those decisions are experienced at local, national, regional and international scales. Therefore, the key question will be how to develop policy to support effective institutional interfaces.

In this section, the authors analyse findings from a larger country-wide study on the implications of policies on creating (or impeding) supportive interfaces. As such, these findings reflect general conditions (not necessarily addressing the key entry points but dealing with country-specific policies) for the formation of supportive interfaces as a basis for future development of policies.

In 2005, the Ministry of Water Resources, National Reform and Development Commission, and Ministry of Civil Affairs in China jointly issued a 'Suggestion on strengthening the establishment of water users associations' (Policy Decision 10), which recommended the establishment of water user associations to manage rural water infrastructure. In response, Baoshan Municipality Water Bureau created 520 water user associations from 2006 to February 2009, covering 142,449 households in 306 villages in 65 townships in the municipality, and managing a total of 13,281 hectares of irrigated land. Based on the suggestion, each county issued its own implementation guidelines to support the establishment of new associations with their own constitutions, elected water user representatives, and regulations governing the operation of the associations, supply of water, repair of infrastructure, collection of fees and management of funds. According to government reports, a supportive interface was realized between the policy implementation and the water user association, which owned and managed its water infrastructure, promoted water saving practices and reduced conflicts in collection of fees.

In contrast to this, some water user associations showed an unsupportive interface due to a lack of funds for their operation, inefficient leadership and a lack of legal clarity regarding their status, which meant that they were operating in legal vacuum. The country-wide study also showed that the ability of the communities to maintain a supportive interface largely depends on the relationships between village leaders and local officials. This is an informal mechanism for obtaining a supportive response from government, and is a barrier to some communities that are not well positioned to obtain the support they require.

Following a significant drought in spring and summer 2005, Longyang district government in Yunnan, China drew up a Drought and Flood Preparation Plan for 2006 and established a drought and flood response co-ordination committee within the water resources bureau. As a member of the committee, Longyang District Agriculture Bureau established an innovative agrometeorology information service called 'Nong Xin Tong', or 'Farmer Information Service' to provide short-term (three- to five-day) forecasts, analysed for agricultural production and sent by text message to subscribers. During a drought in March 2009, the district government sent a 'Notification on strengthening work against the current drought' to all government units mentioned in the plan. In response, Longyang District

Agriculture Bureau established a team to undertake a needs assessment in 18 townships. After the submission of the needs assessment and a recovery report prepared in consultation with the communities, the provincial committee disbursed funds to those townships that had requested relief activities such as pumping machines. The supportive interface between the government plan and implementation was realized, therefore, enhancing the adaptive capacity of the communities.

From a disaster management perspective, it has been argued that policies must be evaluated with respect to economic viability, environmental sustainability, public acceptability and behavioural flexibility (Tol et al., 1996 as cited in Tompkins and Adger, 2003). The same criteria may be extended to evaluate policies addressing climatic stress. Our studies reveal that in general policies have been effective when they receive popular public acceptance.

Government policies such as Indira Awas Yojana in India (on housing) do not take into consideration traditional knowledge and practices, and therefore create an unsupportive interface between the policy intervention and community requirements. In Assam, some groups have a cultural tradition of building their houses on stilts, which increases their resilience in times of flood, while other groups build in styles that make their houses more vulnerable. Although such policies can respond to changing circumstances, it can be difficult to facilitate such change or incorporate effective practices into larger plans because of social constraints to confronting cultural norms. Formal civil society organizations may play a key role in mediating the interface between these informal institutions and broader adaptation planning processes.

The intention of the National Rural Employment Guarantee Act in India is to provide a basic employment guarantee in rural areas to support poverty alleviation and infrastructural development for rural development. The Act was formulated as a result of multi-stakeholder consultations and reserves one-third of jobs for women, providing equal wages. The awareness of these government schemes at a local level provides a supportive interface to enhance the adaptive capacity of the communities, as they generate employment and support infrastructure development. However, most people are unaware of some of the government employment schemes, such as Sampoorna Grameen Rozgar Yojana (SGRY, 2001), and so they are not able to benefit from them.

An analysis of forest policy in Pakistan showed that there is an unsupportive interface between policy implementation and community involvement. The forest policy focuses on conservation of the forest and generating an economical benefit from timber production in consultation with experts who espouse the philosophy of the 'Mother Act of 1894' (FSMP, 2003). The communities around the forest, who depend on forest resources, are considered a threat and are not involved in the planning process, which makes them more vulnerable.

The Alternative Energy Development Policy in Pakistan was developed with broader consultation than most of the other policies, and incorporated the experiences of donor and non-governmental organization (NGO) sectors. This has created a supporting interface between the policy makers and implementers. The research showed that after a few years Chitral valley had over 200 micro-hydro units supported by community-based organizations, donors and NGOs. These initial successes informed the alternative energy policy, and the local renewable electricity initiatives have been completely de-regularized of all the cumbersome requirements of power production and have assured that no one's water rights are violated. To encourage more investment, the government allowed government co-financing and waived all levies and taxes on energy produced through these sources, therefore enhancing the adaptive capacity of the communities.

Labour migration policies in the HKH have different approaches for internal and international migration. It has been found that national policies usually support and facilitate international labour migration and have specific ministries to deal with international migrants. These include the Ministry of Overseas Indian Affairs in India; the Ministry of Overseas Pakistanis in Pakistan; and the Department of Foreign Employment in Nepal. Although the number of internal labour migrants far exceeds that of international labour migrants, these countries do not have specific provisions to deal with internal labour migration. This creates a situation where internal labour migrants become more vulnerable than international labour migrants.

Discussion and conclusion

The success of formal public (normally government) policies and programmes is dependent on their acceptance by formal and informal non-public institutions, including the local communities. Lack of economic benefits and other market drivers often overpower the implementation of government policies, however much they have the capacity to enhance CBA to climatic stresses. The formal and informal institutions interface at key junctures when the policies are being implemented. Although there is no perfect recipe, the success of the policies (and hence the interface) to enhance adaptive capacities has generally been visible when climate context has been considered, local needs and knowledge considered during planning, and local communities have participated in planning.

The field research in the HKH region showed that climate change and variability increase uncertainty and risks in livelihood systems, particularly for those people dependent on agricultural livelihoods. However, markets and government policy (especially in agroforestry in China and Nepal) have a greater impact than climate change awareness in enhancing the adaptive capacity of communities. As policy environments and institutions need to be more responsive to the pace of climate and other drivers of change, they should also be more accountable, more flexible and responsive to the needs of the most vulnerable communities. A proper strategy to translate national and sectoral policies to local-level planning will help facilitate development initiatives to address local adaptation needs. The water governance study in Pakistan showed that local-level institutions and indigenous systems, with support from NGOs, can enhance local adaptive capacities. In the case of flood mitigation structures, it was found that structural measures such as embankments – if supported by non-structural measures including bio-engineering, capacity building and early warning systems – and the participation of local communities enhance a quick response and people's adaptive capacity. Diversifying agricultural production and livelihood systems, including labour migration, are ways in which communities can adapt to both economic and climatic shocks and shifts.

The aim of adaptation planning is to enable and facilitate supportive interfaces where possible (Agrawal, 2010) between formal and informal public institutions and formal and informal non-public institutions. A clearer focus on institutional change in the context of adaptation planning is essential in designing and incentivizing an effective interface between institutions (Dovers and Hezri, 2010). Adaptation cannot occur in a social vacuum – it needs to be supported by institutions and policies designed to enhance the adaptive capacity of local communities.

In this regard, adaptation planning needs to draw on the comparative advantages of institutions that play a role in the development of adaptive capacity, overcoming the tendency of public/private and formal/informal dominion to operate independently of one another. Adaptation planning processes need to be 'bottom up' and capture the services and interventions that people need from the government and other development actors to deal with climatic change and ensure that institutions are in place to deliver them. Regional, national and local adaptation planning needs to identify the institutions delivering services that can contribute to adaptive capacity and put measures in place to ensure that climate information is collected or made available to them and used by them in the design of interventions.

In conclusion, adaptation planning should pay greater attention to the development of effective institutional arrangements to support adaptation, and this requires the assessment of: a) the institutional systems essential to the development of adaptive capacity; b) the interfaces between the institutions within these systems; and c) the factors that can make these interfaces more effective for delivering adaptive capacity.

Acknowledgements

The authors wish to thank the team with which this study was conducted in the HKH region. Overall co-ordination of the study was carried out by ICIMOD with support and guidance from a team from the SEI and the IIED and financial support from the Swedish International Development Cooperation Agency (Sida). The authors thank the local communities and their representatives who participated in this study and the partners in India, Nepal, Pakistan and China who supported this study in their respective countries. Technical support from ICIMOD experts is also acknowledged.

Notes

1 Kunming Institute of Botany (KIB), China.
2 Aga Khan Rural Support Programme (AKRSP), Pakistan.
3 Aaranyak, a society for biodiversity conservation in north-east India.
4 ICIMOD.

References

Agrawal, A. (2006) *Evaluation Report of the Rights and Responsibilities Cooperative Agreement between WRI and USAID*, Ann Arbor, MI: University of Michigan.

Agrawal, A. (2010) 'Local institutions and adaptation to climate change', in Mearns, R. and Norton, A. (eds), *Social Dimensions of Climate Change: Equity and Vulnerability in a Warming World,* Washington, DC: World Bank, pp. 173–98.

Asia Society (2009) *Asia's Next Challenge: Securing the Region's Water Future. A Report by the Leadership Group on Water Security in Asia,* New York, NY: Asia Society.

Bandyopadhyay, J. and Gyawali, D. (1994) 'Himalayan water resources: ecological and political aspects of management', *Mountain Research and Development* 14(1): 1–24 <www.jstor.org/stable/3673735>.

Christoplos, I., Rodriguez, T., Schipper, E.L.F., Narvaez, E.A., Mejia, K.M.B., Buitrago, R., Gomez, L. and Perez, F.J. (2010) 'Learning from recovery after hurricane Mitch', *Disasters* 34(2): S202–19 <http://dx.doi.org/10.1111/j.1467-7717.2010.01154.x>.

Dovers, S.R. and Hezri, A.A. (2010) 'Institutions and policy processes: the means to the ends of adaptation', *Wiley Interdisciplinary Reviews: Climate Change* 1(2): 212–31 <http://dx.doi.org/10.1002/wcc.29>.

FSMP (2003) *Forest Sector Master Plan: National Perspective,* Islamabad: Ministry of Food, Agriculture and Cooperatives, Government of Pakistan.

Hua, O. (2009) 'The Himalayas: water storage under threat', *Sustainable Mountain Development* 56 (Winter): 3c5.

International Centre for Integrated Mountain Development (ICIMOD) (2009) *Local Responses to Too Much and Too Little Water in the Greater Himalayan Region,* Kathmandu, Nepal: ICIMOD.

ICIMOD (2010) *Climate Change Impact and Vulnerability in the Eastern Himalayas: Synthesis Report.* Kathmandu, Nepal: ICIMOD.

IPCC (2007) *Climate Change 2007: The Physical Science Basis,* Contribution of Working Group I to the Fourth Assessment Report of the Intergovernmental Panel on Climate Change, Geneva: IPCC Secretariat.

IPCC (2012) *Managing the Risks of Extreme Events and Disasters to Advance Climate Change Adaptation: A Special Report of Working Groups I and II of the Intergovernmental Panel on Climate Change,* Cambridge, UK, and New York, NY: Cambridge University Press.

Kane, S. and Yohe, G. (eds) (2000) 'Societal adaptation to climate variability and change: an introduction', in Kane and Yohe (eds), *Societal Adaptation to Climate Variability and Change*, pp. 1–4, Dordrecht: Springer Netherlands.

Malone, E.L. (2010) 'The state of societies: vulnerabilities and resilience to the effects of glacier changes', in Malone, E.L., *Changing Glaciers and Hydrology in Asia: Addressing Vulnerabilities to Glacier Melt Impacts,* p. 39, Washington, DC: US Agency for International Development (USAID).

Nellemann, C. and Kaltenborn, B.P. (2009) 'The environmental food crisis in Asia: a "blue revolution" in water efficiency is needed to adapt to Asia's looming water crisis', *Sustainable Mountain Development* 56(Winter): 6–9.

Poteete, A.R. and Ostrom, E. (2004) 'An institutional approach to the study of forest resources', Workshop in Political Theory and Policy Analysis, Indiana University <www.indiana.edu/~workshop/papers/W01I-8.pdf> [accessed 6 October 2013].

Pradhan, N.S., Khadgi, V.R., Schipper, L., Kaur, N. and Geoghegan, T. (2012) *Role of Policy and Institutions in Local Adaptation to Climate Change: Case studies on Responses to Too Much and Too Little Water in the Hindu Kush Himalayas,* Kathmandu, Nepal: ICIMOD.

Rasul, G. (2010) 'The role of the Himalayan mountain systems in food security and agricultural sustainability in South Asia', *International Journal of Rural Management* 6(1): 95–116 <http://dx.doi.org/10.1177/097300521100600105>.

Scheraga, J.D. and Grambsch, A.E. (1998) 'Risks, opportunities, and adaptation to climate change', *Climate Research* 11(1): 85–95 <http://dx.doi.org/10.3354/cr011085>.

Tompkins, E.L. and Adger, W.N. (2003) *Defining Response Capacity to Enhance Climate Change Policy,* Tyndall Centre Working Paper No. 39, Norwich: Tyndall Centre for Climate Change Research and School of Environmental Sciences, University of East Anglia.

Xu, J., Grumbine, R.E., Shrestha, A., Eriksson, M., Yang, X., Wang, Y. and Wilkes, A. (2009) 'The melting Himalayas: cascading effects of climate change on water, biodiversity, and livelihoods', *Conservation Biology* 23(3): 520–30 <http://dx.doi.org/10.1111/j.1523-1739.2009.01237.x>.

About the authors

Neera Shrestha Pradhan is a Water and Adaptation Specialist at ICIMOD. Ms Pradhan has more than 13 years of experience in the field of environmental management, freshwater conservation and management, climate change adaptation and community adaptation. She is a civil engineer with master's degree in environmental management and has worked with the International Union for Conservation of Nature and WWF prior to ICIMOD.

Vijay Khadgi is Assistant Co-ordinator at ICIMOD and has worked for nearly 10 years in the fields of hydrology, disaster management and climate change adaptation. He has developed several communication strategies on these topics and has a profound interest in conducting research in these areas.

Nanki Kaur is a senior researcher with the climate change group at the International Institute for Environment and Development. She has over 10 years of experience in applied research for policy development in the area of climate resilient development planning, natural resource management, pro-poor development and aid effectiveness.

CHAPTER 7

Economic analysis of a community-based adaptation project in Sudan

Muyeye Chambwera and Khitma Mohammed

This chapter presents the costs, and their redistribution, of scaling up climate change adaptation. Using a community project addressing current climate-related stresses in Kassala state of Sudan as a proxy for an adaptation pilot, it assesses whether pilot project activities contributed to adaptive capacity and if they could be eligible for scaling up, and how costs would be redistributed among players and when scaling up to the entire state. The pilot indeed contributed to the building of adaptive capacity, and thus is eligible for scaling up. In the pilot, a non-governmental organization (NGO) and government covered 75 per cent and 24 per cent of project costs respectively. On scaling up the pilot, the costs would be redistributed so that the role of government, private sector and communities in meeting costs would increase, while that of NGOs would go down. Also, focus and costs would shift towards functions that enable communities to adapt autonomously at scale.

Keywords: climate change adaptation, economics, costs, drought, adaptive capacity, scaling up, Sudan

Climate change is increasingly recognized as a critical challenge to human well-being and development. Variations and extremes of climate disrupt food production and water supply, reduce incomes and damage properties. Adaptation can reduce these impacts. At the same time, if development is properly planned, it will help build adaptive capacity. The economics of adaptation have taken on increasing importance because of the likely high costs of adaptation, which will compete with existing priorities for resources. Adaptation to climate change presents itself as an economic problem because it addresses the bigger problem of allocating scarce resources to attain sustainable development (Chambwera and Stage, 2010). Very little economics work has been done in developing countries (Margulis et al., 2008) where competition for resources between adaptation and other needs is high. Community-based adaptation (CBA) especially has received little economic analysis, partly because it is an emerging field.

The few adaptation pilots that are under way and the development projects that address existing climate variability provide a useful starting point for planning larger-scale adaptation. Assessment of the cost of upscaling small projects helps decision makers to plan, prioritize and integrate adaptation into

http://dx.doi.org/10.3362/9781780447902.007

their development plans at national and sectoral levels by building on what already exists.

This chapter is based on a case study undertaken in Kassala state of Sudan, whose main focus was to estimate the cost of scaling up adaptation pilots, and to map out a pathway for upscaling the project by assessing the distribution of the costs of adaptation among different players. The case study specifically analysed the contribution to climate change adaptation of a project implemented by a non-governmental organization (NGO) in Kassala state from 2006 to 2009. We refer to it in this chapter as the 'pilot' project, as we assume that it contributes to climate change adaptation by building local capacity to address climate-related stresses. The specific objectives of the case study were to:

- conduct a climate change vulnerability assessment in the project area, which forms the basis for adaptation planning;
- assess the project's successful contribution to adaptation, which forms the basis for scaling it up;
- assess the cost of scaling up adaptation at the community level in Kassala state, including the distribution of costs among different players.

Kassala state

Geography

Kassala state is located in the eastern part of Sudan and covers an area of 42,282 square kilometres. The state shares borders with Eritrea to the east, Nile River state in the north, Gezira state to the west and Gedarf state in the south.

More than 80 per cent of the area of Kassala state comprises flat plains while rocky highlands and hills constitute the rest of the area. Heavy clay soil covers most of the land in the southern parts of the state, where the most important irrigated and rain-fed agricultural projects are located.

The state has a total population of around 1.8 million with an annual growth rate of 2.5 per cent and average family size of 6.2 people. Dependency rates within families are very high; for instance, 50 per cent of the population is either below 16 or over 70 years of age. A high percentage of the state's population lives in rural areas. Most of the population is agro-pastoralists. The rural sedentary population is estimated at 63 per cent of the total Kassala state population while 12 per cent are pastoralists.

Climate

Kassala is divided into two climatic zones: arid and semi-arid. Its climate is characterized by low irregular rainfall from June to September, persistent drought (with a long dry season), land degradation and desertification. The winter season is characterized by dry wind from December to February.

Temperature. The temperature ranges between 33 and 41 degrees Celsius. Mean daily temperature is fairly high, with the maximum in the hottest months above 40 degrees; these temperatures are usually recorded in May. The minimum average temperature is 16 degrees, usually recorded in January. The wind is always north–south.

Rainfall. The rainfall is characterized by significant variations in distribution as well as in timing and location, with precipitation ranging between 83 mm per year in the extreme north to 300 mm in the extreme south, in dry and semi-dry ecological zones respectively. However, there is a noticeable long-term decrease in rainfall (Zakieldeen, 2009). Rainfall for the years 1996 to 1998 showed a marked decline in the mean, indicating the drought conditions prevailing in the state. The climate of the northern parts of the state is also affected by the Red Sea.

Climatic events. Drought, the major hazard, is frequent, leading to food insecurity and devastating consequences in most parts of the state. The state has experienced extreme climate events since the 1980s, which result in food deficits and a high dependency on food relief.

Flood is another climatic hazard in Kassala. For instance, the flooding of the Gash river in 2003 and 2007 affected a large number of people and resulted in destruction of their assets and their basic savings.

Economic profile

Despite recent progress, per capita income is low and infrastructure remains inadequate. Most rural households depend on agriculture-related activities. Long-term problems depressing the state economy include:

- illiteracy and lack of access to information;
- prevalence of epidemic diseases;
- increase in the number of displaced people and refugees;
- natural disasters such as flooding and drought;
- lack of financial resources;
- decrease in per capita incomes (Kuwait Fund for Arab Economic Development, 2010).

Main economic activities. Kassala is an important agricultural centre and a source of border trade to the country. However, the domestic economy of the state is subject to fluctuations in the flow of goods and is highly sensitive to conflict within and outside the borders. The state's economy depends largely on the traditional activities associated with natural resources. The potential for using natural pastures for crop cultivation has been greatly affected by armed conflict in the region. Most farmers and pastoralists abandon their lands for urban areas.

- *Cultivation.* The total area of arable (cultivable) land in Kassala state is about 1.67 million hectares, which is equivalent to 40.5 per cent of the total area of the state. The actual cultivated area ranges between 0.46 million and 0.66 million hectares, is mostly rain-fed and involves 60 per cent of farmers. However, agricultural production in the rain-fed area is not more than 16 per cent of the total production of the same area under irrigation. The main crops grown in the state are millet, sorghum, maize, cotton fruits and oil seeds (groundnut and sesame).
- *Livestock.* Livestock rearing in Kassala state is considered one of the dominant economic activities, practised by almost 80 per cent of the rural population. Livestock production systems are predominantly traditional. Livestock reared include cattle, sheep, goats, poultry and pigs. The livestock capacity is estimated to be 6.41 million head, of which about 830,000 are cattle, 2.93 million sheep, 1.9 million goats and 750,000 camels, on which about 190,000 pastoralists depend for their living. The contribution of livestock to gross domestic product is 21 per cent (Ministry of Animal Resources and Fisheries, 2009).

  The pastoral production system in Kassala is adopting a set of strategies that facilitate survival by utilizing different ecosystems with multiple resources. Most of the tribal groups have started to change from being purely nomadic to semi-nomadic and agro-pastoralist as a result of the recurrent drought.
- *Tourism.* Kassala's natural environment provides a great opportunity for tourism, but this potential is yet to be utilized, and would require a considerable level of investment if tourism were to become a meaningful economic activity in the state.
- *Industry.* The only industry present is a sugar factory and a crafts industry.

Overall, conflict and displacement in the state, accompanied by harsh waves of drought, impose high levels of pressure on the available resources. As a result, land available for grazing and rain-fed agriculture has decreased, leading to food insecurity and many tribal conflicts over resources. About 19 per cent of families in the state do not meet their food requirements. Food insecurity has significantly increased during the last decades, particularly in the rural areas.

The pilot project

The pilot project 'Resilience to Poverty of Rural Communities in Kassala State' aimed to contribute to strategies for reduction of long-term poverty and vulnerability in rural communities, and was implemented in 18 villages of Kassala state, targeting 15,963 people (2,661 farmers, 5,407 pastoralists, 5,340 internally displaced people and 2,555 women).

Specifically, the pilot project aimed to sustain and improve the viability of traditional small-scale rural production systems to underpin rural livelihood security through three broad activity areas:

- *Activity 1.* Strengthening the capacities of local producer groups to enable them to plan and manage their own livelihood development strategies using available resources and to influence policy change.
- *Activity 2.* Increasing their ability to cope with drought through improved access to water harvesting and agricultural technologies.
- *Activity 3.* Reducing conflict over natural resources through improvement of natural resource management and conflict resolution skills among farmers, pastoralists and government.

The targeted project area depends on agricultural production mainly for subsistence and has suffered from constraints imposed by drought as well as by local markets and the policy environment. The area is vulnerable to droughts and there are severe water shortages in summer. This is where most of the internally displaced persons and many pastoral groups are settled. The pastoralists and nomads are used to spending summer in the rangeland on the two sides of the Gash river, and so the area is prone to conflicts between farmers and pastoralists. The area close to the river is subject to severe soil erosion.

Local project partners were community-based organizations (CBOs), development agencies and local government departments with direct links to project activities, specifically the Ministry of Agriculture, Forestry, Irrigation and Livestock, and the Ministry of Animal Resources and Fisheries in Kassala state.

Case study research methods

The case study used an approach that enables the identification and quantification of the roles played by different stakeholders at different levels in CBA, including their contributions to building local adaptive capacity through different ways and channels. A simple model showing adaptation investments and channels was developed (Table 7.1). The main variables of the model are:

- *Assets and entitlements.* These include social, physical, natural, human and financial assets that enable the stakeholders to respond to climate change.
- *Process.* These are activities through which value is created, such as production, processing, marketing and management.
- *Flows of goods and services.* These include inputs, information, energy and maintenance. These goods and services are important for the activities to take place.

- *Enabling conditions.* These are factors outside the immediate control of local communities, but that influence their decisions to invest in adaptation, such as national policies, land tenure, prices, subsidies, etc.
- *Stakeholders.* These are formal and informal, individuals or organized groups, and include households, businesses, local and national government, NGOs, CBOs, etc.

All stakeholders in the system have a direct or indirect interest in building local adaptive capacity, and may also be sources of adaptation resources, or a channel for adaptation investment. The level of investment (cost) required to scale up adaptation is evaluated by its source, target, implementation level and channel, as shown in Table 7.1.

Table 7.1 Simple model for evaluating adaptation by source, level and channel

Adaptation function category	*Stakeholders*					
	Households	*Informal businesses*	*Formal businesses*	*NGOs*	*Local government*	*National government*
	Investment value (cash or kind)					
Assets and entitlements						
Processes and flows						
Enabling conditions						
Total gross						

Case study research first undertook a vulnerability assessment of the project area, focusing on key assets, followed by an assessment of the roles played by different actors in the pilot project. Primary data were collected through questionnaires, semi-structured interviews and group discussions in March 2010. Secondary data were collected from the literature and pilot project documents of the lead NGO partner. Random sampling techniques were used to select the interviewees from different villages. Seven out of the 18 villages in the project were selected for the questionnaire survey and focus group discussions.

Stakeholders were fully involved in the evaluation of the interventions. These stakeholders included the institutions that directly service the target groups, such as the Ministry of Agriculture, Forestry, Irrigation and Livestock (Director of Pastureland and Fodder, Director of the Department of Planning), Ministry of Animal Resources and Fisheries (Director of the Department of Planning and Projects), Sudanese Standards and Metrology Organization, Women's Development Association, village development committees (VDCs), Elgandoul Association, Pastoralist Union, Small Producers Union, Department of Planning and Development, the monitoring and evaluation section at the Ministry of Finance, and NGOs operating at the state level, such as the International Fund for Agricultural Development (IFAD).

Assessment of vulnerability

Kassala state is one of the most vulnerable areas in Sudan. Frequent drought cycles remain the main climatic factor challenging the sustainable livelihoods of the local communities. Drought is a common phenomenon on Kassala's rangelands; it has become an annual occurrence in the state. People in the affected areas have lived with drought over a long period of time. The most severe droughts in Kassala state occurred during the years 1965, 1970 and 1984. However, the frequency of occurrence is higher in recent times than in the past. The biggest events from the late 1980s onwards occurred from 1988 to 2000–02, and recently between the 2006–07 and 2007–08 cropping seasons.

The local people asserted the prevalence of unfavourable conditions for farming owing to the sharp decline in rainfall. This has led to declines in agricultural productivity, food insecurity and loss of herds, and has limited their sources of income. Analyses show that farmers' perceptions of climate change are in line with the climatic data records.

Assets most vulnerable to drought

The assets that are heavily affected by drought include agricultural lands (production and productivity), rangelands, water resources, soils and livestock. Frequent droughts and other contributing factors such as overgrazing, deforestation and poor agricultural practices have resulted in the degradation of soils and rangelands and decreased yields on traditionally cultivated soils. The soil has been eroded and has become infertile. Rangeland cover (area) and pasture quality (species composition) have declined. According to the Department of Rangelands, the drought of 2009 resulted in a decrease in rangelands of approximately 40 per cent of the total area of pasture; according to our analyses, this translates to about $25 million.

The respondents in the case study noted that the 2009 drought and water scarcity led to livestock loss and reduced household farm productivity and income, resulting in people migrating to other areas. They indicated that as a result of drought the area under cultivation decreased to less than half while productivity also decreased. According to our analyses, this loss is equivalent to $33.75 million. Thus the combined loss due to impacts on rangelands and cropping lands in 2009 was $58.75 million. According to one of the farmers in the case study: 'Due to the drought, the area we used to cultivate [1.2–4.1 hectares per farmer] decreased to less than half and its productivity decreased by approximately 90 per cent.'

Considerable numbers of livestock have been lost as a result of recurrent droughts. Although the livestock populations are increasing significantly because of improved veterinary and drinking water services, a high rate of mortality of livestock took place during drought years, as shown by data from the Department of Planning and Projects in the Ministry of Animal Resources and Fisheries (Table 7.2).

Table 7.2 Animal losses in drought years over and above losses in normal years

Livestock type	*1984–85 (% of loss)*	*1990–91 (% of loss)*
Cattle	20	10
Sheep	40	15
Goats	10	5
Camels	5	2.5

Source: Ahmed et al., 2004

Coping strategies

Farmers adopted a range of coping strategies to minimize the adverse impacts of drought in order to maintain their livelihoods. They responded by changing crop varieties and sowing dates, planting different crops, harvesting water (constructing terraces), selling livestock and taking on casual work. The pastoral groups responded to the new constraints with several strategies including diversification of herds, increasing mobility, reliance on crop residues for their animals, changing to agro-pastoralism and selling livestock at low prices. Famines and the scarcity of drinking water forced pastoralists to abandon their villages and migrate to urban centres for alternative livelihoods. They sell animals when in need of cash for agriculture, or use a failed crop to make fodder for livestock. Pastoralists move depending on their own forecast of environmental change. Some coping mechanisms are destructive, such as overgrazing, wood cutting and making charcoal for sale. The various coping measures need to be supported by investments in watershed management, drought-resistant crop varieties, farm diversification and capacity building. The vulnerability analysis indicates that the community is vulnerable to drought, which has an impact on the livelihoods of the people and their income-generating activities. Natural assets are heavily affected by the recurrent drought. Many coping mechanisms have been adopted by small-scale farmers and pastoralists, but these mechanisms are exhausted in the face of the recurrent droughts.

Costing the pilot project

The pilot project sought to reduce the vulnerabilities discussed above with activities that sustain and improve the viability of small-scale rural production systems to underpin rural livelihood security, with capacity for communities to develop and implement their own livelihood development strategies, and to increase resilience to cope with drought. The pilot project cost a total of $1,378,000 (€984,000) over a period of three years (2006 to 2009).

Multiple stakeholders were involved in the implementation of the pilot project. An assessment of the contributions of different stakeholders shows that the pilot was mainly financed by external donors through an NGO (up to 75 per cent of total costs), while households made contributions of less than 1 per cent of the project costs. Table 7.3 summarizes the direct contributions of different stakeholders towards assets, processes and enabling conditions. Most resources were channelled into developing assets and entitlements which are critical for building long-term resilience and adaptive capacity.

Table 7.3 The cost of the pilot project to different stakeholders

Adaptation function category	*Stakeholders*			
	Government[2] US$ (%)	*Donors/NGOs US$ (%)*	*CBOs US$ (%)*	*Households US$ (%)*
Assets and entitlements				
Including investment in natural, physical, social, human and financial assets[1]	199,839 (15.00)	821,155 (60.00)	13,000 (0.94)	1,000 (0.07)
Processes and flows				
Crop production, food processing, animal husbandry, provision of energy and information, and other inputs	129,546 (9.40)	212,000 (15.40)	390 (0.03)	610 (0.04)
Enabling conditions				
Policies (land tenure policy) and enforcement	–	–	–	–
Total	*329,385 (23.90)*	*1,033,155 (75.00)*	*13,390 (0.97)*	*1,610 (0.11)*

1 Such as training in water harvesting technology, natural resource management, formulation of community networks, establishment of centres for CBOs, seed stores, terracing, provision of credits, etc.

2 'Government' includes Ministry of Agriculture, Forestry, Irrigation and Livestock, Ministry of Animal Resources and Fisheries, and Ministry of Environment.

Note: The government contribution was in kind (technical contribution), which has been valued in monetary terms.

Assessment of the project's contribution to adaptive capacity

Key assets supported

Although the project was designed to benefit 15,963 people as direct beneficiaries, it reached around 6,885 of the poorest households (42,687 people) in 28 villages in rural Kassala who are dependent on traditional small-scale farming. The high level of participation in the project implementation of target communities across their different social and ethnic groups, through their VDCs, zone development committees and their umbrella association (Elgandoul Association), as well as the good contribution made by government institutions in the implementation and monitoring of the project activities, led to high achievement of the project's outputs and outcomes.

Feedback from stakeholders in the project area shows that the three main outcomes generated by the project were capacity building of local producer groups, increased resilience to cope with drought, and resolution of conflicts over natural resources. During group discussions, all participants revealed that the project has had clear impacts on their livelihoods. Farmers are aware of the technical approaches to cope with the changing environment. The resilience of food production systems in the face of climate change improved through pilot adaptation measures in demonstration sites while livelihood opportunities were diversified. As a result of the project, an animal resource forum was also established. This forum comprises government institutions at national and state levels and NGOs. It aims to protect the state herds from climate threats, train herders on climate change, attract financial support, bridge the fodder gap, improve livestock marketing outlets, organize pastoralists in assemblies or groups, and provide credits and veterinary services, demarcation of routes, etc.

The project has also contributed to reducing the severe effects of observed climatic changes on subsistence rain-fed farming, animal husbandry and related livelihood activities in rural communities of the state. It also contributed positively to the maintenance and conservation of the rural environment and to the resolution of conflicts between farmers and pastoralists over natural resources through demarcation of animal routes.

One of the major impacts of the project was the creation of a spirit of co-operation between the community and the government that saw them both involved in the development and implementation of project activities.

Impact of the project on key assets

Social assets. The capacities of local producers were strengthened, and they became better able to express their own needs, prioritize their urgent problems, and plan and implement their own livelihood development strategies. They also established working relationships with other institutions and were able to influence agriculture-related policies. This has been achieved through the creation of CBOs and VDCs and an umbrella organization (Elgandoul Association) and by promoting environmentally friendly activities, training, information sharing, advocacy, mobilization, lobbying, networking activities and formal registration. The CBOs and VDCs formed under the project have good links with other institutions, facilitating access to agricultural inputs.

Participants claim that before the project they were not aware of climate change. With their awareness now raised, members of VDCs and CBOs and their umbrella association can lead negotiations with local government. The link that was created between different rural women's groups and the Women's Development Association empowered women to perform their own development activities and facilitated their access to the market for selling their products.

Human assets. The project improved human capacity by establishing committees and equipping local producers and traditional leaders with an awareness of strategic planning, management, leadership, negotiation skills, book keeping, fundraising, food processing and new technologies. This resulted in them strengthening their capacity to develop their own strategies to cope with droughts.

Training in traditional terrace construction

Training in modern terracing methods

Physical assets. The project provided farmers with material help: improved agricultural practices, water management, and other agricultural resources such as agricultural extension services and veterinary services; and established development centres. As a result, the capacities of small producers were strengthened and their resilience to cope with drought increased.

The project also supported the construction of a small dam to provide water for livestock and other uses using new technology; this is cited as one of the major project achievements by the government and local communities, and is recommended for replication.

Natural assets. Land and water management have improved land productivity. Natural resource conservation and the green culture among target groups improved as a result of training on energy efficiency, tree planting and natural resource management.

Financial assets. A revolving fund was established for the distribution of animal drugs and liquefied petroleum gas cooking stoves, which were distributed to women-headed households. This resulted in improved livelihoods. Farmers also started to generate some income from vegetables and cash crops.

Training in home garden cultivation

Flows. The project resulted in significant productivity improvements in the rural communities of Kassala state. It introduced donkey ploughs that can plough an area of one feddan (0.42 hectares) in four working hours; it normally takes two to three days to cultivate using hand tools. Productivity increased from two sacks to seven sacks per feddan. Terracing led to an increase in food production in the area despite the low rainfall and long dry spell. In the 2007 season, the area under cultivation increased by 18 per cent and yield increased by 10 per cent. Farmers also produced sorghum for their food and fodder for their animals, in addition to vegetable cash crops that were introduced by the project and from which they could generate income.

An example of a functioning home garden

Improved sorghum in cultivation

Harvest of the improved sorghum

The project interventions assisted women to diversify their livelihoods by generating some income from agricultural production as well as providing extra food for their own consumption. The women who received food processing training also helped to disseminate the learned techniques to other women not directly involved. The project has directly contributed to reducing the vulnerability of local communities to climate change and variability and to building their adaptive capacity. It has had immediate benefits in people's livelihoods and natural ecosystems. Its direct contribution to the resilience of local communities has been achieved by building on their experiences and indigenous knowledge. Farming experience, livelihood diversification, access to credit, access to water, tenure rights, off-farm activities and access to extension and animal health services are the main factors that enhance the adaptive capacity of local communities.

As this project demonstrates, building climate change adaptive capacity may involve investment in different areas, some not necessarily directly related to climate change; for instance, the project enabled farmers to market their crops and livestock. Enabling conditions are also required, including policies and legal frameworks such as those relating to land tenure, to avoid conflicts between farmers and pastoralists. In order to increase the value of the project under review, other needs include climate and price information, the construction of roads, agricultural extension, and veterinary and research services. This investment should take place at different levels, including national, state and community levels. This confirms Schelling's (1992) conclusion that good development is one of the best forms of adaptation and the postulation by Burton et al. (2006) that the success of human adaptation to climate changes will depend heavily on development options and choices, and that a higher level of development is likely to produce greater adaptive capacity.

Scaling up the pilot: redistribution of costs

The pilot project contributes towards Sudan's National Adaptation Programme of Action. Thus, the scaling up of this project to cover the wider community, building on the experiences and lessons learned, can contribute to Sudan's wider adaptation goals. However, in this project, NGOs covered the bulk of the costs, which will not be the case when activities are scaled across the entire state. Thus the active involvement of other players will be key, requiring a redistribution of financial responsibilities and other contributions at different levels.

Relevant agencies at the **national level** include policy and planning authorities that function at the scale of an entire country (in this case, the Ministry of Finance, Ministry of Investment, Ministry of Agriculture, Ministry of Irrigation, research institutions, and Higher Council for Environment and Natural Resources). Their role is to ensure that climate change adaptation is embedded in development policies, and to allocate resources to adaptation in the national budget. They also lead on the external mobilization of funds for adaptation.

Policy and planning authorities within relevant sectors at **state level** have a direct lead in project proposals and the translation of national-level priorities and budgetary allocation into sectoral and local development plans, factoring in adaptation.

At the **local level**, local government, local communities, NGOs, CBOs and the private sector should be involved in identifying and prioritizing their needs and putting plans into action. All actors at all levels should invest in adaptation functions, including activities, assets, flows and enabling factors, as this helps in sustaining adaptation actions.

To estimate how the costs of adaptation could be redistributed when upscaling this pilot project to the entire state, group discussions, individual stakeholder consultations and an analysis of the state budget allocated to activities that contribute to related activities were used during the case study research. The stakeholders suggested that additional ways and players will be necessary to make scaling up sustainable. The private sector and the issuing of loans to local communities were suggested as additional ways of redistributing the costs of adaptation when scaling up this pilot. Stakeholders noted that the pilot project did not involve agricultural companies, especially those that supply inputs, or local businesses that are critical for the distribution of inputs and other assets needed for adaptation. With income-generating activities showing potential during the pilot period, local enterprise development could be stimulated through the provision of loans. The experience from revolving funds under the pilot showed the potential of this approach.

The total cost of scaling up the pilot project to the entire state was estimated at $11.4 million per year, with the cost being met by several players. The analysis shows that government will make the largest contribution to the costs of adaptation (52 per cent), followed by the private sector (17 per cent). NGO support (13 per cent) and loans (13 per cent) are also seen as significant in meeting the costs of adaptation, while the direct contribution of households is expected to be 5 per cent of the total costs. The support to production processes will require the highest share of resources (66 per cent), while assets would need 18 per cent of the total budget and policy reform 16 per cent. Table 7.4 gives a breakdown of costs.

Table 7.4 Contribution of different actors to the cost of adaptation to scale up the project

	Stakeholders				
Adaptation function category	*Government US$ '000 (%)*	*Community[1] US$ '000 (%)*	*NGOs US$ '000 (%)*	*Private sector[2] US$ '000 (%)*	*Loans US$ '000 (%)*
Assets and entitlements					
Including investment in natural, physical, social, human and financial assets[3]	925.3 (8.1)	228.2 (2.0)	329.4 (2.9)	369.0 (3.2)	129.6 (1.1)
Processes and flows[4]					
Crop production, food processing, provision of agricultural services and inputs, information	3,106.6 (27.3)	322.8 (2.8)	1,179.6 (10.3)	1,579.4 (13.9)	1,355.6 (11.9)
Enabling conditions					
Policies (land tenure policy)	1,872.2 (16.4)	–	–	–	–
Total (%)	*51.8*	*4.8*	*13.2*	*17.1*	*13.0*

1 Includes households and CBOs (such as farmers' and pastoralists' unions).
2 Such as agricultural companies and businesspeople.
3 Such as training in water harvesting technology, natural resource management, formulation of community networks, construction of terracing, construction of animal routes, establishment of centres for CBOs, seed stores, provision of credits, etc.
4 Includes management, technology, agricultural processing and production.

As the cost estimations above are based on the pilot project under review, they reflect both the costs of facilitating community actions and the costs of providing material inputs where required. The consultations on the costs of upscaling did not look at the detailed unit costs, as these are assumed to be those from the pilot project, but focused on how these costs could be redistributed across different players. Thus the actual inputs may differ from community to community.

Policy implications and conclusions

The contribution of pilot projects to building key community and household assets, improving their production processes and improving the enabling environment is part of the evidence of their contribution to building local adaptive capacity, while providing for the involvement of a wider set of stakeholders helps spread the costs of adaptation. Even though the pilot project in Kassala state was not conceived as an adaptation project, it made significant contributions towards climate change adaptation through the various assets and processes, and provides a very good learning platform for climate change adaptation.

NGOs are important, and could take the lead in implementing and financing early adaptation and pilots; however, after the pilot phase, government and local communities will assume greater responsibility for the long-term financing and implementation of adaptation projects. It is important for external financiers to channel their contributions appropriately. The scaling-up budget of $11.4 million does not necessarily need to be channelled through government, and its contribution is only about half the total cost. Other channels need to be considered as well, such as working with NGOs.

Involving the private sector and providing communities with loans to engage in livelihood diversification and income-generating activities could reduce the amount of donor and government funding when scaling up pilots. This opportunity is still limited in Kassala state, where most of the production, especially in agriculture, is for subsistence consumption, and experience with enterprises is very low.

The analytical approach used in this case study could be developed further and used as one of the tools for planning CBA as it provides rich information on costs of adaptation and their distribution, and therefore who should pay. The analytical approach could also be used for targeting external and government support to help only those who are most vulnerable to climate change, such as the poor and vulnerable members of the community.

References

Ahmed, M.M.M., Idris, M.F.E. and Teka, T. (2004) *Dryland Husbandry in the Sudan: Grassroots Experience and Development*, Dryland Husbandry Project (DHP) Sudan, Addis Ababa and Khartoum: Organisation for Social Science Research in Eastern and Southern Africa (OSSREA) and University of Khartoum.

Burton, I., Diringer, E. and Smith, J. (2006) *Adaptation to Climate Change: International Policy Options*, Washington, DC: Pew Center on Global Climate Change.

Chambwera, M. and Stage, J. (2010) *Climate Change Adaptation in Developing Countries: Issues and Perspectives for Economic Analysis*, Environmental Economics Discussion Paper, London: International Institute for Environment and Development (IIED).

Government of Sudan (2009) *Kassala State Situational Analysis*, Kassala state: GoS.

Kuwait Fund for Arab Economic Development (2010) 'Kassala state profile', *The International Donors and Investors Conference for Eastern Sudan* [online] <www.kuwait-fund.org/eastsudanconference/index.php?option=com_content&task=view&id=128&Itemid=294&lang=english> [accessed 16 December 2013].

Margulis, S.M., Bucher, A., Corderi, D. et al. (2008) 'The economics of adaptation to climate change: methodology report', Washington, DC: World Bank <http://siteresources.worldbank.org/INTCC/Resources/MethodologyReport0209.pdf> [accessed 9 November 2013].

Ministry of Animal Resources and Fisheries (2009) 'Animal resource forum', Kassala state: GoS.

Schelling, T. (1992) 'Some economics of global warming', *American Economic Review* 82(1): 1–14 <www.jstor.org/stable/2117599>.

Zakieldeen S.A. (2009) 'Adaptation to climate change: a vulnerability assessment for Sudan', Gatekeeper series no. 142, London: IIED <www.acts.or.ke/institute/docs/climate_sudan.pdf> [accessed 16 December 2013].

About the authors

Dr Muyeye Chambwera is a technical adviser with the United Nations Development Programme in Botswana. His work focuses on sustainable development. He has previously worked for the International Institute of Environment and Development in London, leading the theme on the economics of climate change, developing and applying economic methods for climate change adaptation that are suitable for developing countries.

Dr Khitma Mohammed is an economist currently working as senior researcher for the Higher Council for Environment and Natural Resources in Sudan's Ministry of Environment and Physical Development.

CHAPTER 8

Growing rooibos and a stronger community: participation and transformation

Bettina Koelle and Katinka Waagsaether

This case study focuses on a community of small-scale farmers based in the semi-arid, western interior of South Africa who harvest the indigenous rooibos tea plant. They have joined forces to enhance their livelihoods and to learn about fair trade markets, climate change and community development in order to address poverty and historical injustices. In 2000 the Heiveld Co-operative was founded as a vehicle for processing and marketing the farmers' rooibos tea and at the same time supporting the community's learning journey. A climate change adaptation process commenced in 2004 and has become an important vehicle for community and personal development. This case study describes the process of learning and exploration carefully facilitated by two non-governmental organizations (NGOs). Critical reflections focus on the skills required by the service providers and facilitators of such learning processes, how the lessons learned in South Africa could be extended more globally, and prerequisites and possible processes to move from adaptation towards transformation to a more resilient community.

Keywords: participation, transformation, participatory action research, learning, co-operative, South Africa

Adaptation, transformation and participatory action

Adaptation to climate change has become a topic closely linked to development and justice debates. Adaptation of human systems to climate change is defined by the Intergovernmental Panel on Climate Change (IPCC) as 'the process of adjustment to actual or expected climate and its effects, in order to moderate harm or exploit beneficial opportunities' (IPCC, 2012).

This definition can be linked to global efforts by many international and national organizations to support the most marginalized groups in anticipating change – as they are more likely to be susceptible to the expected impacts of climate change (Adger et al., 2006; Adger, Paavola and Huq, 2006; Tschakert and Dietrich, 2010; Van den Berg and Feinstein, 2009).

http://dx.doi.org/10.3362/9781780447902.008

Adger et al. summarize the link between justice and adaptation as follows:

> Vulnerable groups are also likely to be at the sharp end of the policy response to climate change. They have not been given a choice as to whether they would like to adapt to climate change. They face increased risks yet are ignored when policy decisions regarding mitigation and adaptation are made. We live in a world where future climate injustices are likely to compound past injustices, such as underdevelopment and colonialism, that themselves have resulted in the uneven patterns in today's world. (Adger, Paavola and Huq, 2006)

Pelling (2011) differentiates three different types of adaptation processes:

1. *Adaptation as resilience* – returning to the status quo or maintaining the status quo: The systems remain intact or recover from stresses but do not challenge the status quo. Considering that frequently the pre-existing systems do not represent a desirable state for many poor and marginalized groups, this approach does not necessarily address social or environmental injustices.
2. *Adaptation as transition* – some change: The system undergoes some change and is in transition towards another state. Adaptation in this context means that the existing systems are questioned and that incremental change is set in motion.
3. *Adaptation as transformation* – transformative change: Reconfiguring the structures of development, this approach aims at changing the overarching political-economic regime. Profound transformation allows redistribution of security and opportunity in society.

Participatory action research (PAR) is an approach that aims at a transformative process (Kemmis and McTaggart, 2000), and thus the PAR process can be closely linked to adaptation processes. PAR is an approach by which research and development are conducted from within, rather than on, practice. The crux of why it is favourable for research and development to be conducted from within is illustrated by Max-Neef (1991). Using the example of love, Max-Neef highlights how one can read and study everything that is written on the phenomenon of love, but that one can never truly understand love without falling in love. Accordingly, he argues: 'If we have so far been unable to eradicate poverty, it is because we know too much about it, without understanding the essence of its existence as well as the mechanism of its origins' (Max-Neef, 1991: 102). The PAR process, working from within, allows adaptation practitioners and researchers to be co-learners and co-researchers in the groups with which they are working. In their exploration of the field of PAR, Kemmis and McTaggart (2000) find that, while there is no unitary approach, PAR can be seen as:

- an ethical approach that acknowledges co-responsibility for the outcomes of actions;

- a social process that explores the relationship between the realms of the individual and the social;
- emancipatory, in that it aims to release people from the constraints of irrational, unproductive, unjust and unsatisfying social structures that limit their self-development and self-determination;
- critical, in that 'it is a process in which people deliberately set out to contest and to reconstitute irrational, unproductive, unjust and/or unsatisfying ways of interpreting the world, ways of working and ways of relating to others';
- reflexive, in that it aims to help people investigate reality in order to change it.

It is thus an approach that aims to transform both theory and practice. Kemmis and McTaggart (2000) also note that the PAR process tends to involve a spiral of self-reflective cycles of: planning a change; acting and observing the process and consequence of the change; reflecting on these processes and consequences; and then re-planning; acting and observing; reflecting; and so on, as illustrated in Figure 8.1.

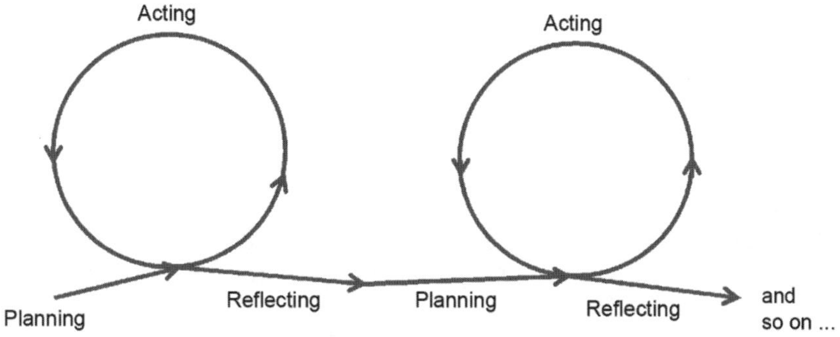

Figure 8.1 The self-reflective learning cycle in participatory action research processes

An important aspect of the PAR process is that it is, ideally, context specific, and thus shaped by the specific situation in which it is applied. The importance of context-specific approaches is illustrated in the argument made by Max-Neef in his book *Human Scale Development* (1991: 19), that while fundamental human needs are 'finite, few and classifiable', the satisfiers that allow for the satisfaction of those needs are culturally determined and highly diverse. Accordingly, the transformative PAR process of self-reflective learning cycles can be strengthened when taking into account the multiple needs and the ways in which these are best satisfied in the local context. Participants in PAR processes should therefore also carefully consider the type of satisfiers that they promote and support. Max-Neef (1991) groups satisfiers into a number of generic categories, of which **synergic satisfiers**, those that fully satisfy a given need while simultaneously contributing to the satisfaction of other needs, are most desirable. Less desirable are the **pseudo satisfiers**, elements that give rise to a false sense of satisfaction of a given need, and that undermine the possibility of fully satisfying the need that they were aimed at fulfilling. An example of a synergic satisfier is popular education, which addresses the need for understanding. On the other hand, chauvinistic nationalism could be described as a pseudo satisfier of the need for identity (Max-Neef, 1991). Accordingly, development and adaptation processes should promote awareness of the multiplicity of needs and the importance of applying synergic satisfiers in development and adaptation processes (Max-Neef, 1991).

Understanding the needs and satisfiers at the level of a community or society provides important insights into the focus of a possible transformative adaptation process that is able to address the potential 'pathologies' in the existing system. PAR approaches are well suited to supporting such a process of transformation in practice.

The study area: the Suid Bokkeveld plateau

The area of the Suid Bokkeveld was originally inhabited by San people – hunter-gatherers who were the first people of South Africa. Today their rock art remains as a silent witness to this complex culture and its people. More than a millennium ago, Khoi pastoralists arrived in the area, adding greater complexity to the cultures and resource use there. Following the settlements of European colonists in the Suid Bokkeveld after 1739, and the subsequent centuries of colonialism, oppression under the apartheid regime and the modernization of society, the San and Khoi cultures of the Bokkeveld have largely vanished. Today the inhabitants have lost the indigenous local languages and have Afrikaans as their first language. San and Khoi identity have also largely been lost (Oettle, 2005; Oettle et al., 2009).

The Suid Bokkeveld was marginal farming land and thus not proclaimed under the Apartheid Group Area Act, which segregated racial groups spatially in allotted areas. Following the dismantling of apartheid and the advent of democratic governance in South Africa, land ownership patterns and wealth distribution have not changed significantly. As a result, the social pattern that evolved over an extended period of time in the Suid Bokkeveld has remained relatively intact. The farms are scattered, and the distances between farms and homesteads vary from 2 kilometres to 30 kilometres. There is no public transport and communication is challenging (Oettle et al., 2009).

The Suid Bokkeveld lies within the semi-arid winter rainfall region of the western interior of South Africa. The area receives most of its rainfall between April and September, with some sporadic rains in the summer period from September to March.

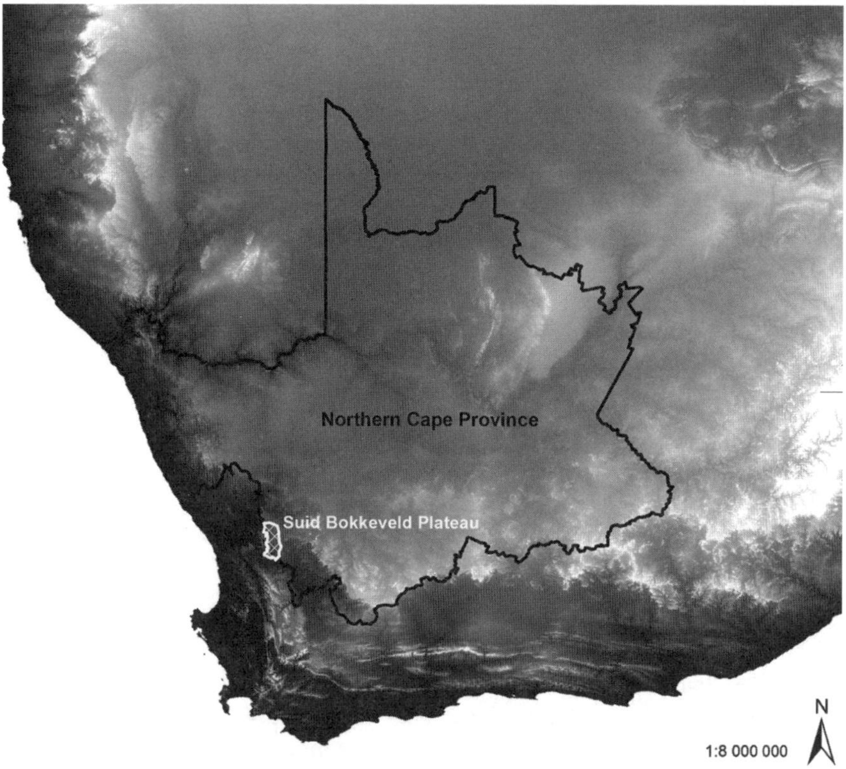

Figure 8.2 Northern Cape Province (South Africa) and the Bokkeveld
Credit: B. Koelle

The participatory action research process in the Suid Bokkeveld

Although agricultural lands of the Suid Bokkeveld are of a marginal nature, two global biodiversity hotspots are located in the area: the Cape Floral Region (CFR) and the Succulent Karoo. Rooibos tea (*Aspalathus linearis* – an endemic plant of the CFR) has recently become increasingly popular on local and global markets. Due to its ability to grow productively in low rainfall areas on infertile lands without irrigation, rooibos has become an economic boon in the CFR. In order to directly benefit from the economic opportunities provided by rooibos, group of previously disadvantaged small-scale farmers founded the Heiveld Co-operative in 2000 to ensure access to organic and fair trade markets worldwide (Oettle et al., 2009). Since its humble beginnings, with an initial membership of 14 and initial capitalization of 1,400 rand ($143), the Heiveld Co-operative has matured as a successful business with a participatory, democratic structure through which its 60 member farmers have derived economic benefits as well as being able to communicate and collaborate more effectively with one another (Oettle, 2012).

The learning and participatory action research process started in 1998 when two South African NGOs, Indigo development & change (Indigo) and the Environmental Monitoring Group (EMG), attended two farmers' meetings with officials of the Northern Cape Department of Agriculture, in the course of which participants expressed their frustration regarding the lack of effective support in their development processes, poor access to resources and conflict within the community. At a subsequent community meeting in 1999, residents expressed their frustration at the extremely low prices they received for their rooibos harvests (in some cases less than the harvesting costs to the farmers) and the lack of employment opportunities in the area. The farmers and NGOs agreed to work together, with the following principles guiding their collaboration:

- Involvement in any project activity should include contribution and benefit.
- People's vision, enthusiasm and contribution should be mobilized before benefits are achieved.
- The least advantaged should benefit the most.
- The project should benefit both the local community and the wider community.
- Everybody undertakes to work together in the spirit of mutual respect.
- There should be transparency regarding all project documentation.

A two-day participatory workshop followed, including visioning exercises and problem and objective trees to better understand the complexity of the situation and the dreams and hopes of the people in the Suid Bokkeveld. A PAR approach, 'which seeks to enhance people's ability to learn together in the course of taking action to improve their situation', was chosen for the road ahead. It was further agreed that a 'PAR is underpinned by a strong belief that change for the better must be driven by those whose lives are to be improved and not by "outsiders"' (Report of the Suid Bokkeveld Community Workshop, 27–28 March 1999).

Following the initial community workshop in the Suid Bokkeveld, Indigo and EMG facilitated a community exchange visit to the Wupperthal Co-operative and a commercial organic rooibos farmer in order to create opportunities for inspiration and learning. Farmers were able to witness at first hand the advantages and challenges of rooibos production and marketing. On their return home, the farmers participating in the exchange visit decided to form a collective organization focusing on joint production and marketing of their produce: rooibos tea. In 2000, the Heiveld Co-operative Ltd was founded – a co-operative buying and processing the rooibos tea of its members and selling this to the fair trade and organic market (Oettle, 2005; Oettle and Koelle, 2003; Oettle et al., 2009).

Over the course of the following years, Indigo and EMG were able to provide a level of ongoing support to the Heiveld Co-operative, and had regular contact with community members through workshops and meetings. These interactions were largely interlinked, and crises and conflict, related to

The participatory community workshop in the Suid Bokkeveld in 1999. *Credit:* B. Koelle

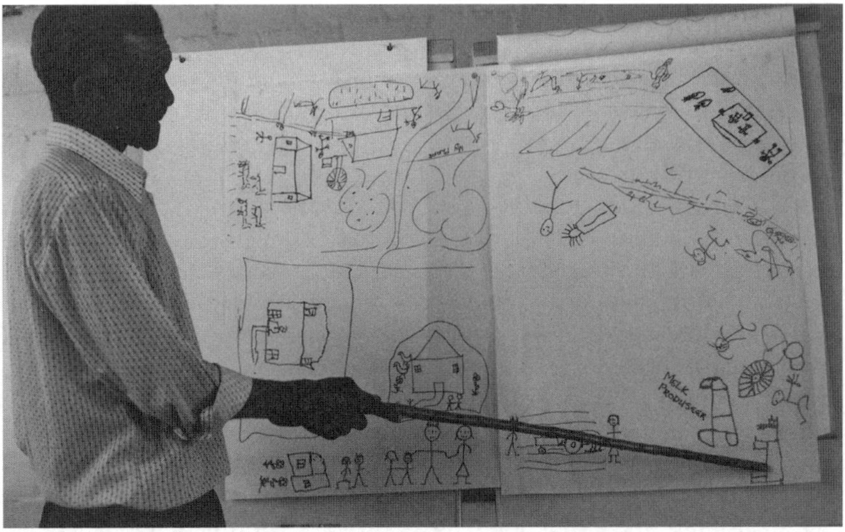

A farmer presents the vision of his working group for the Suid Bokkeveld. *Credit:* B. Koelle

the co-operative and to other matters, were dealt with while learning how to better solve problems and how to anticipate them. Through this continuous interaction and co-operation, based on mutual understanding and respect, the need for specific research or capacity development would arise, leading to the establishment of a number of projects and initiatives.

One of the important initiatives that materialized, and which has now become a golden thread of interaction between the community members and the NGO practitioners, is quarterly climate change preparedness workshops. The workshops were initiated in 2004, at a time when farmers were experiencing the first year of the most severe drought in living memory. Almost a decade later, the workshops are still taking place every three months and have become a cherished local institution. Here, the farmers share experiences relating to the weather events of the previous months, and also reflect on happenings that have taken place in their farming community.

The NGO facilitators are responsible for the workshop logistics, such as transport and refreshments, and they also prepare an agenda for each workshop based on a format that has evolved through years of interactions between participants and facilitators. Regular agenda items include: reflection on the weather events of the last quarter; presentation of the seasonal forecast for the coming quarter; report back from ongoing PAR processes and other research; and discussion on general community issues. Each workshop also includes innovative aspects that respond to emerging situations, requests and opportunities.

In this regard, it is also crucial for the NGO facilitators to practise flexibility, letting go of pre-established workshop agendas to focus on the issues that the participants deem important that day. Therefore, the workshops also serve as a forum for discussion of internal political issues, often leading to the planning of actions not necessarily related to the topics under debate at the workshop. Giving participants the ability to continuously shape the workshop agenda has created a crucial sense of ownership and mutual respect, and grounds the process within the PAR approach, with research and development being sparked from within.

Some of the other initiatives and projects arising from the continuous interaction between the NGOs and the farmers from the Suid Bokkeveld are listed in Table 8.1. These are all individual PAR processes that have emerged through the overarching PAR process of interaction and co-operation. The projects and initiatives were suggested by farmers and planned and implemented in collaboration with farmers, local NGOs and scientists as appropriate. Table 8.1 shows a summary of PAR topics undertaken in the past decade.

The application of the PAR process aligned to the principles defined in Kemmis and McTaggart (2000), and described above, has involved co-generation of research questions, research action and knowledge, and thus co-responsibility and a common sense of ownership for the action outcomes. An important outcome of this co-generation and co-responsibility is the mutual respect and power balance that it has created between farmers and practitioners. Such aspects can often be seen in small things and acts, and a good way to illustrate this is to describe a small incident that might at first seem insignificant. Some years ago, when farmers started monitoring their local weather, this was done using forms on sheets of paper. These observations

Table 8.1 Action research processes in the Suid Bokkeveld since 2000

Year started	*Research topic*	*Farmer researchers*	*Collaborating researchers*	*Status*
2003	Sustainable harvesting of rooibos (wild and cultivated)	Male farmers	UCT EMG	Concluded
2006	Germination of wild rooibos	Male farmers and one woman farmer	UCT USB EMG Indigo	Ongoing
2008	Water monitoring on farms	Women farmers (and some male farmers)	UCT Indigo CGS	Ongoing, with a special focus on women farmers
2008	Erosion control and organic farming	Men and women farmers	Heiveld EMG	Ongoing
2010	Climate monitoring and climate diaries	Two men and three women farmers	Indigo UCT	Ongoing
2012	Heat stress and livestock: thresholds	Women farmers	CSIR UCT EMG Indigo	Ongoing

Note: The partners involved were: University of Cape Town (UCT), University of Stellenbosch (USB), Environmental Monitoring Group (EMG), Indigo development & change (Indigo), Heiveld Co-operative Ltd (Heiveld), Council for Scientific and Industrial Research (CSIR), and Council for Geoscience South Africa (CGS).

were discussed at each climate change preparedness workshop, and the sheets of paper were then stored in a database by the NGO practitioners. This was the accepted process until one of the farmers turned up at the Indigo offices asking to have his observation sheets, because, he explained, he wanted to conduct his own research. This illustrates the strong ownership of the process felt by farmers, and that the process of collecting data sheets clearly was not appropriate. In collaboration with the farmer, a 'climate diary' was designed, and has since been used for registering weather observations and monitoring data. The climate diary belongs to the farmers, and Indigo keeps records of the data, if agreed upon by the farmer, by taking pictures of the observations recorded in the diary.

Other characteristics of a PAR process can be illustrated in the interventions that relate to the organization of the Heiveld Co-operative, as it works to release people from the constraints of their social and economic structures. The Heiveld Co-operative provides farmers with a means by which they can access previously inaccessible markets and unattainable prices, thus transforming the limiting structures of a fractured, isolated small-scale farming community. More in-depth aspects of this transformation are outlined below.

The Heiveld Co-operative: a vehicle for social transformation

In the light of people's complex livelihoods and global change, the Heiveld Co-operative has become an important agent of change – not just economically but, and perhaps more importantly, it has helped build the self-confidence and bolster the courage of individuals to engage in new learning processes. The Heiveld Co-operative has evolved as a key organization to promote social transformation: a dynamic local organization owned and managed by the farmers themselves.

The initiation and management of the organization is therefore important not only for the economic benefits for its members, but also to revive and sustain a significant joint identity that is key in supporting the learning process. As is the case with all agricultural businesses, the co-operative and its members have continuously had to keep learning and innovating, responding to changes and challenges, both internal and external (Oettle, 2012). The structure and operations of the co-operative have thus shaped members' ability to handle conflict and to promote rights, especially those of vulnerable groups within the co-operative (such as women and other marginalized individuals and groups). This is illustrated in the co-operative's effort to support children and youth in the Suid Bokkeveld, with fair trade premiums having been used for the provision of bursaries for tertiary studies and for financial support to improve the services provided to local learners (Oettle, 2012). The Heiveld Co-operative also has a specific gender policy – encouraging women to become members in their own right and to serve on the Heiveld management structures, and encouraging the employment of women. All three permanent members of staff are women.

The Heiveld Co-operative has also actively expanded its networks over the past years, offering opportunities for members to participate in national and global processes (such as the United Nations Framework Convention on Climate Change 17th Conference of the Parties (UNFCCC COP17), the BioFach organic food trade show in Germany, fair trade processes, Terra Madre gatherings of the Slow Food movement in Italy, etc.). These active networks have had a strong transformative effect on the Suid Bokkeveld community – even on those not able to undertake journeys themselves. This can be illustrated by a woman farmer, seeing a photograph of her daughter presenting at COP17 in the daily digest: 'I was so proud of her – she was doing so well – I cried. She was there for the entire Suid Bokkeveld.' Indigo has focused on engaging women in the various activities, and on creating spaces where women feel safe to stand up and speak. Using participatory video as a tool, a movie reflecting on the impact of development processes in the community in the past 10 years was compiled (Parring, 2013). In the so-called 'Indigo movie', a woman farmer of the Suid Bokkeveld highlighted: 'When people visit other places they come back and give feedback. I learned a lot from my cousins who have travelled to workshops and conferences. Like Lena who went, I feel I can do this too.'

The learning and innovation journey of the Heiveld Co-operative has sparked both intended and unintended synergy effects, as illustrated in Table 8.2.

Table 8.2 Services and synergy effects provided by the Heiveld Co-operative

Services of the Heiveld Co-operative	*Beneficiaries*	*Synergy effects*
Joint production facility	Heiveld members	Better market prices, predictable facilities and pricing, pride in members' ownership of assets, employment, venue for learning processes such as climate change preparedness workshops
Access to short-term production loans	Heiveld members	Financial predictability for families, ongoing productivity, perceived financial security, identity
Development programmes for the most marginalized groups	Most marginalized groups in Suid Bokkeveld	Feeling of community and solidarity, supporting marginalized groups, allowing women to enter the production of rooibos tea through production support
Market access to global organic and fair trade markets	Heiveld members	Heiveld members are invited by trading partners and visit Italy, France, Switzerland, Germany, Belgium, etc. on various occasions, expanding networks, understanding of global marketing conditions, pride in the community
Promoting solar energy and green business	Suid Bokkeveld community	In collaboration with local NGOs and clients of the Heiveld, solar energy programmes have been rolled out in the entire Suid Bokkeveld
Supporting interests of members and the larger fair trade movement	Heiveld members, global fair trade movement, small-scale farmers worldwide	Linking to global networks, inspiration through interaction with other farmers, agencies, partners
Engaging in community-based adaptation	Suid Bokkeveld farmers, global adaptation community	Learning process supporting increased self-confidence, experimentation strengthens networks with partners, sharing learning and formulating adaptation strategies enhances anticipatory capacity, sharing process methodology with a global audience, learning from other global and local processes

Elements of the learning process: synergies for adaptation?

The learning process has been ongoing and has evolved over the years, involving different individuals with various research questions. Initially, more mature men engaged most actively in the learning journey, but soon more and more women joined the learning processes. With support from the Heiveld Co-operative, the learning process now includes a wide range of people, from old to young, men and women.

The learning processes have not been linear or centrally orchestrated. Some learning processes were project-related, others emerged and were self-organized, and some were linked to the Heiveld Co-operative and the organizational development process.

Some interesting aspects of this learning process are described below. Together, these threads make up a rich tapestry of learning.

Community monitors: supporting ownership of research processes

Scientists and practitioners were working with community members in the PAR processes. Some of the so-called community monitors are women and men pursuing a defined research question, often in collaboration with other community members and/or scientists. Community monitors are paid a stipend for their work since their engagement prevents them from taking on other income-generating activities. The research design is undertaken collaboratively and regular feedback meetings are held with fellow researchers and the wider community to discuss the implications of findings for current practice and to agree upon the way forward. Frequently, the community monitors inspire other community members to explore their own research questions. This element has contributed a broad sense of ownership to the research processes, but has also encouraged the farmers to explore their own questions in collaboration with scientists and other resource people – thus becoming the driver of their own investigation and learning processes. As men and women have been involved as community monitors, this has led to an interesting shift in gender balance over the years (Koelle, 2013).

Learning journey and gender: specific needs and interests

As noted above, the 'learning journey' was initially dominated by men. Men in the Suid Bokkeveld are often more mobile, have access to transport and are more used to interacting with a wider community. Women were often housebound and obliged to look after the children, especially at weekends. The specific design for women's research programmes (designed with and for women) has overcome some of these barriers. When workshops are planned, parallel workshops are organized for children, so that women are free to attend the adult workshops with peace of mind. This particular innovation increased the participation of women from virtually zero to 60 per cent. The learning journey could thus accommodate men and women, recognizing specific needs and interest expressed.

Climate diaries and seasonal forecasts: linking to climate science

The interaction of farmers with seasonal forecasts is crucial in understanding and engaging with the anticipated variability of the climate. However, it has also proved important that farmers themselves monitor the environment to better understand long-term and slow change and to better communicate with scientists. Manual climate monitoring (using a maximum/minimum thermometer and a climate diary) and automated weather stations are used in a synergic manner. The seasonal forecasts are interrogated in different ways, including through the seasonal forecast game played at the quarterly climate change preparedness workshops.

Climate change preparedness workshop: platform for synergic learning

These workshops take place every three months and are a key element of the joint learning process. While the workshops are social gatherings aiming at a creative learning process, they are also a place to share information, to see children presenting to the adults, to feel supportive and proud (Archer et al., 2008). The importance of satisfying a range of needs within one workshop process was applied more comprehensively over time, resulting in the workshop participant numbers increasing dramatically. The workshops are the golden thread of the learning process and include aspects of sharing, learning, exploring and planning. They also often contain practical experiential learning events that support sustainable farming practices.

Sustainable farming: farmers teaching farmers

Mentor farmers are employed by the Heiveld Co-operative especially to support sustainable organic farming practices among members. The mentor farmers also advise on and monitor the sustainable harvesting of the wild rooibos tea and advise in the planning and implementation of farming activities and restoration measures. They also advise on adaptation measures and link to the climate change preparedness workshops.

Extreme weather events and adaptation strategies

Increasingly extreme events have been observed by members in the community. Because of this, adaptation strategies are planned jointly in order to anticipate both these events and long-term change. This anticipatory planning requires input from all the learning processes mentioned here.

The learning process has been a multifaceted one, and its richness has included interactions with visitors from South Africa and internationally, as well as the farmers themselves sharing their learning and experiences at conferences and other forums (such as the UNFCCC COP).

The learning process is supported by different individuals in the community. In different ways they are contributing and benefiting. The learning process spans from the individual level to the organizational level, therefore allowing a more synergic perspective. However, the process of creating these synergies for learning remains a challenge in complex community adaptation processes.

Facilitation of participatory action research processes for transformation

Facilitating PAR processes can be challenging and requires a certain skill set and methodology. From our practice, a few key points emerged.

Appropriate process

The PAR process must be specific to the situation, the culture and the individuals involved. There is no 'blueprint' for the learning process. Most important is that the process is led by the intended beneficiaries and not by donors, NGOs or researchers with agendas not visible or aligned to community interests. In the Suid Bokkeveld, the evaluation exercise at the end of each learning event opens the space for critical reflection on the learning process and lets new ideas emerge. This has become an important ritual and is a good way to shape the learning process in partnership.

Skills of facilitators

In order to facilitate these learning processes, a facilitator should be courageous and flexible at the same time, while having excellent listening skills. The facilitator is an enabler and not a leader; understanding this is a great step forward towards designing and facilitating sound participatory processes. These skills are not necessarily taught at university level and thus often have to be acquired in the field or on training courses offered by NGOs or the corporate sector.

Gender issues

Indigo has focused on engaging women in the various activities, and on creating spaces where women feel safe to stand up and speak. As was highlighted by a woman farmer in the community in the Indigo movie:

> Indigo offered courses and training workshops. We went, and this was the first time we had this opportunity. Us women now have more opportunities. Often, it is mainly men who engage in these processes. But now a lot of opportunities have been created for women to participate.

This has empowered women to stand on their own feet and, as was highlighted by another female farmer in the Indigo movie:

> Through the climate change preparedness workshops, I learned when to plant and when not to plant. It really means a lot to me as a woman. Now I don't have to ask the men all the time when to plant. Now I think for myself and I feel very proud of that.

Often we assume that vulnerable groups are homogeneous and we overlook the more marginalized groups within a 'community'. It is therefore always important to understand the situation within the community and to collect gender disaggregated data to improve our understanding of gender dynamics in adaptation.

Moving from adaptation to transformation: the way forward

The example of the learning journey in the Suid Bokkeveld has illustrated some aspects of a very complex and colourful process. This process has not been linear and has not been driven by a specific project, but rather has been a rich collection of action, research and learning by different actors over the past 10 years.

The question remains: did this learning journey lead to a process that Pelling (2011) would describe as adaptation for transformation in the Suid Bokkeveld?

Comparing the situation today with the status quo 12 years ago, some significant change can be observed:

- Initially there was no community organization supporting rooibos farming or marketing. The Heiveld Co-operative, established during the PAR process, has been an important vehicle for organizational learning and has enabled the farmers to create and sustain a profitable market for their products while contributing towards gender equity in the community and sustainable organic farming practices through training.
- While women were largely absent in the early stages of the community process, they are now strong partners in various PAR processes, actively contributing to local, national and international workshops and events.
- While before 2000 scientific research was inaccessible to the small-scale farmers, research is now co-ordinated through a research protocol, requiring researchers to share their results and demonstrate a community benefit. This has changed the power balance between farmers and scientists and has led to a number of collaborative action research processes involving scientists, farmers and practitioners.
- The farmers in the Suid Bokkeveld were very isolated in 2000 but over the past decade they have managed to create a national and international network of partners they can draw on for advice and assistance. These networks make the farmers more resourceful in addressing crises or in preparing for extreme weather events.
- Young learners, initially not part of the community learning process, are now actively engaging in community workshops and exploring self-determined learning journeys.
- Fractures in the community prior to 2000 were very visible and were an obvious obstacle to economic and social development. This has shifted to a perception of increased opportunities for farming, learning and solidarity clearly emerging in community workshops and Heiveld Co-operative management structures.

Despite these achievements, we do not argue that transformation for adaptation has been achieved. It would take more radical and longer-term interventions to create a profound system shift in the community and to address the social and environmental injustices of the past. The case study cannot demonstrate a radical system change. However, a level of transition has been achieved through the PAR learning process over the past 10 years, and the aspirations for a more profound change have been kindled. This case study therefore shares some insight into the complex process that has changed some aspects of individual and community learning. There is reason to hope for a profound system shift and a more just society in the Suid Bokkeveld and further afield.

References

Adger, N., Paavola, J., Huq, S. and Mace, M.J. (eds) (2006) *Fairness in Adaptation to Climate Change,* Cambridge, MA, and London: MIT Press.

Adger, N., Paavola, J. and Huq, S. (2006) 'Toward justice in adaptation to climate change', in Adger, Paavola, Huq and Mace (eds), *Fairness in Adaptation to Climate Change,* Cambridge, MA, and London: MIT Press.

Archer, E.R.M., Oettle, N.M., Louw, R. and Tadros, M.A. (2008) '"Farming on the edge" in arid western South Africa: climate change and agriculture in marginal environments', *Geography* 93: 98–107.

IPCC (2012) 'A special report of Working Groups I and II of the Intergovernmental Panel on Climate Change: summary for policymakers', in Field, C.B., Barros, V., Stocker, T.F., Qin, D., Dokken, D.J., Ebi, K.L., Mastrandrea, M.D., Mach, K.J., Plattner, G.-K., Allen, S.K., Tignor, M. and Midgley P.M. (eds), *Managing the Risks of Extreme Events and Disasters to Advance Climate Change Adaptation,* Cambridge, UK, and New York, NY: Cambridge University Press.

Kemmis, S. and McTaggart, R. (2000) 'Participatory action research', in Denzin, D. and Lincoln, Y. (eds), *Handbook of Qualitative Research,* London: Sage.

Koelle, B.R.I. (2013) 'Women farmer scientistsin participatory action research processes for adaptation', in Alston, M. and Whittenbury, K. (eds), *Research, Action and Policy: Adressing the Gendered impacts of Climate Change,* Dordrecht: Springer.

Max-Neef, M.A. (1991) *Human Scale Development: Conception, Application and further reflections,* New York, NY: The Apex Press.

Oettle, N.M. (ed.) (2005) *Enhancing Sustainable Livelihoods in the Suid Bokkeveld,* Amsterdam: Both Ends.

Oettle, N.M. (2012) *Adaptation with a Human Face: A Case Study,* Cape Town: Environmental Monitoring Group.

Oettle, N. and Koelle, B. (2003) *Capitalising on Local Knowledge: Community Knowledge Exchange,* Washington, DC: World Bank.

Oettle, N., Goldberg, K. and Koelle, B. (2009) 'The Heiveld Co-operative: a vehicle for sustainable local development', Drynet Case Study, Cape Town.

Parring, S. (2013) The Indigo Movie. 15 min. Nieuwoudtville: Indigo development & change.

Pelling, M. (2011) *Adaptation to Climate Change: From Resilience to Transformation,* London and New York, NY: Routledge.

Tschakert, P. and Dietrich, K.A. (2010) 'Anticipatory learning for climate change adaptation and resilience', *Ecology and Society* 15(2): 11 <www.ecologyandsociety.org/vol15/iss2/art11/>.

Van den Berg, R.D. and Feinstein, O.N. (2009) *Evaluating Climate Change and Development,* New Brunswick, NJ: Transaction Publishers.

About the authors

Bettina Koelle is a geographer and has been working for Indigo development & change (a South African NGO) for the past 10 years, focusing on the links between community development, natural resource management and climate change adaptation. She has been working with rural vulnerable and marginalized communities in southern and eastern Africa using participatory methods. She has a master's degree from the Free University of Berlin and is currently studying towards her PhD at the University of Cape Town.

Katinka Lund Waagsaether has been working with Indigo development & change for the past two years. Her focus areas include co-ordination of the South African Adaptation Network and engagement in the national and international climate change adaptation landscape, as well as co-ordination of Indigo's grassroots projects. She has a master's degree in environmental and geographical science from the University of Cape Town, where her research focused on the vulnerability of small-scale farmers in the rural north-eastern parts of South Africa.

CHAPTER 9

Strengthening the Food for Assets approach for community adaptation in Makueni, Kenya

Victor A. Orindi, Daniel Mbuvi and Joel Mutiso

Over 80 per cent of Kenya is classified as arid and semi-arid lands, experiencing frequent droughts with occasional flash floods. Innovative ways are needed to enable communities to recover and accumulate assets for increased resilience in the face of increasingly variable climate. This case study highlights the impacts of one such strategy – Food for Assets (FFA) – as used in Makueni county, in supporting adaptive capacity at the local level. It shows how the FFA approach has enabled households to strengthen their assets and invest in alternative livelihood options, through support for enhanced rainwater harvesting and soil conservation efforts. Emerging lessons are that forging partnerships is key in ensuring the sustainability of community-based adaptation initiatives; that asset accumulation is necessary for community resilience; and that having an exit strategy right from the beginning encourages ownership.

Keywords: community-based adaptation, food for assets, adaptive capacity, collaboration, resilience, arid and semi-arid lands, Kenya

Kenya is highly vulnerable to climate change. Over 80 per cent of the country falls under arid or semi-arid land (ASAL) categories, characterized by high temperatures and low precipitation that is unevenly distributed in space and time. These factors, together with historically inappropriate policies and development approaches, have undermined the major livelihood strategies of the rural population based on crop farming and livestock keeping. The increasing frequency of extreme events – droughts and, to a lesser extent, flash floods – is undermining the resilience of households and communities to recover from shocks and to cope with future climate change when productive assets are lost beyond viable thresholds. Research by Herrero et al. (2010) shows that Kenya experiences major droughts every decade and minor ones every three to four years. This is too frequent to allow poor households to recover fully from such impacts. This chapter will show how the use of the Food for Assets (FFA) strategy, though not designed specifically to address climate change, can help towards the twin objectives of achieving food security and increasing adaptive capacity at the local level. Using the Africa Climate Change Resilience Alliance's local adaptive capacity (LAC) framework

http://dx.doi.org/10.3362/9781780447902.009

(Jones et al., 2010), we have looked at how the FFA approach currently contributes to building adaptive capacity and how this may be strengthened going forward.

Food for Assets is defined as an integrated community development strategy involving the use of food aid, labour and participatory decision-making approaches in order to develop productive assets that are owned, managed and maintained by households or communities (C-SAFE, 2004). FFA as an approach aims at increasing and preserving productive assets among targeted vulnerable households and communities who happen to be food insecure in most cases. Projects implemented under FFA are those that can be started and completed by local communities without the need for external labour or technical assistance, with the exception of tools. FFA differs from Food for Work in that it emphasizes the creation of assets to a given group, which is a more long-term process. The fact that FFA is aimed at increasing resilience by creating assets through participatory decision making largely qualifies it to be a community-based adaptation (CBA) strategy. In addition, assets created are diverse and to a large extent strengthen ecosystem resilience, as in the case of soil and water conservation structures.

There are areas of weakness in the FFA approach: the food aid component has been blamed for increasing dependency syndrome among beneficiaries and therefore killing innovation, which is key for enhancing adaptive capacity. Additionally, the roles of the institutions involved need to be audited, and access to and use of information on climate change and markets enhanced. These areas of weakness need to be addressed if FFA is to achieve its intended objective of improving food security and strengthening the livelihood assets that are necessary for adaptation under an increasingly variable climate and uncertain future.

In the following sections we outline the LAC framework that we have used to analyse the FFA strategy in terms of how it has been used in Makueni, Kenya to manage climate-related risk, in particular the recurring drought. We focus on types of assets created, institutions involved and level of participation in decision making, resource mobilization and benefit sharing, innovation, knowledge and information sharing – which are the key areas of emphasis for strengthening adaptive capacity at the local level.

The local adaptive capacity framework

According to Jones et al. (2010), the LAC framework identifies five distinct but interrelated characteristics that are conducive to adaptive capacity: asset base, institutions and entitlements, knowledge and information, innovation, and flexible forward-looking decision making. Table 9.1 outlines features of each of these characteristics that reflect high adaptive capacity.

Table 9.1 The local adaptive capacity framework's five characteristics and their features

Characteristic	*Features that reflect a high adaptive capacity*
Asset base	Availability of key assets that allow the system to respond to evolving circumstances
Institutions and entitlements	Existence of an appropriate and evolving institutional environment that allows fair access and entitlements to key assets and capitals
Knowledge and information	Ability to collect, analyse and disseminate knowledge and information in support of adaptation activities
Innovation	Ability to create an enabling environment to foster innovation and experimentation and the ability to explore niche solutions in order to take advantage of new opportunities
Flexible forward-looking decision making and governance	Ability to anticipate, incorporate and respond to changes with regards to the system's governance structures and future planning

Source: Jones et al., 2010

We have used the LAC framework because it focuses on the local level, and incorporates intangible and dynamic dimensions of adaptive capacity together with capital and resource-based components in its analysis of adaptive capacity (Adger et al., 2003; Eriksen and Kelly, 2007; Jones et al., 2010). We reviewed the FFA strategy as it is relatively new in this part of Kenya (as discussed below), and has great potential to strengthen adaptive capacity among poor and vulnerable communities. In the sections that follow, we look at the FFA strategy through the LAC lens.

The history of Food for Assets in Kenya

In order to strengthen the resilience of smallholder farmers to drought, the Government of Kenya (GoK) supported the design and implementation of the second phase of the Arid Lands Resource Management Project (ALRMP)[1] between 2003 and 2010. One of the key components of this initiative was an innovative FFA project approach, supported by the United Nations World Food Programme (UNWFP). The FFA aims at enhancing asset creation, investment in alternative livelihoods and community capacity building on technologies for mitigation against the adverse effects of drought.

Since the 1980s, the GoK, together with UNWFP, has been providing free food to populations affected by droughts in order to address food and nutrition insecurity. This is necessary to save lives. However, food aid may also create a dependency syndrome among beneficiaries and can act as a disincentive to engage in productive activities (Orindi and Ochieng, 2005). Free food is costly and may kill innovation and ultimately lower the resilience of poor and vulnerable groups. These weaknesses with food aid have led to a shift to FFA in recent years, with more focus on creating and protecting households and community assets to enhance recovery from droughts and build resilience to shocks.

The first phase of FFA activities in Kenya began in October 2010 and targeted 15 ASAL districts. It was co-ordinated at the national level by an FFA secretariat anchored in the Ministry of State for Development of Northern Kenya and Other Arid Lands and at the district level by district FFA co-ordinators. The activities are implemented by nine partner agencies: World Vision Kenya, Kenya Red Cross Society (KRCS), Turkana Rehabilitation Project, Child Fund Kenya, Garissa Rehabilitation Project, the Catholic diocese of Meru, the Catholic diocese of Kitui, Action Aid Kenya and the Consortium of Cooperating Partners. Implementation follows guidelines in a manual produced by GoK and WFP (2010).

The FFA approach has been used in many countries in sub-Saharan Africa, including Ethiopia, Malawi, Zambia and Zimbabwe (C-SAFE, 2004), with mixed success. The common objective is that it aims to meet immediate food needs as well as building assets to strengthen resilience among targeted households and communities. In addition to the assets created, other long-lasting effects of FFA include the skills built to plan and manage micro-initiatives. A good example is capacity building on rainwater harvesting structures: layout, implementation and operation, as well as maintenance of the completed structures, are achieved through a continuous engagement of GoK technical extension officers to provide technical support and to train the beneficiaries from project inception to completion.

In Kenya, FFA projects seek to align themselves with the *National Climate Change Response Strategy* of 2010 (GoK, 2010) and the sessional paper in 2012 on the sustainable development of ASALs, which calls for better co-ordination among institutions involved in climate adaptation and development in ASALs in order to use scarce resources efficiently and to avoid unintentionally undermining each other's work. Indeed, poor or no co-ordination among the many players in ASALs is partly being blamed for the maladaptation associated with past projects or interventions in these areas (Gok, 2012).

FFA projects are generally expected to result in the following outcomes:

- improved pasture and browse production;
- improved diversification of food sources;
- improved access to water for both human and livestock consumption;
- reduced environmental degradation;
- improved access to markets and other sources of food (through feeder roads).

Masongaleni location in Makueni county

Masongaleni is situated in the semi-arid county of Makueni in the Eastern Province of Kenya. The map in Figure 9.1 shows the major livelihood zones in the county.

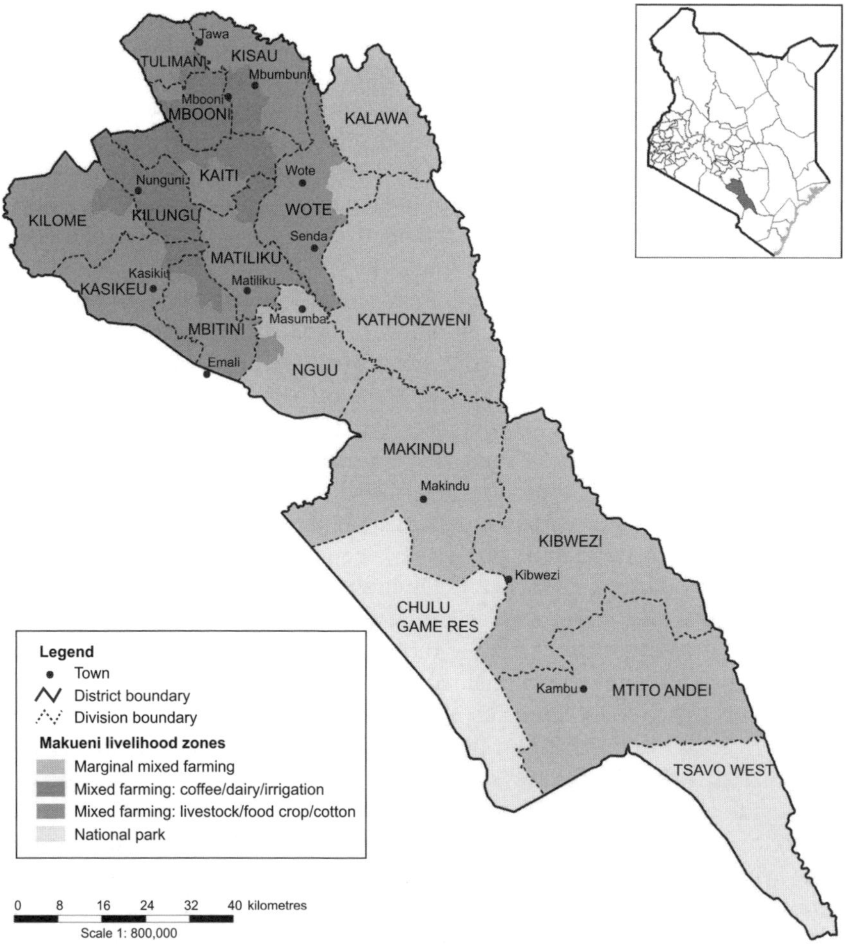

The boundaries and names shown and designations used on this map do not imply official endorsement or acceptance by the United Nations

Figure 9.1 Livelihood zones in Makueni county

Source: UNWFP, Kenya Vulnerabilities and Mapping Unit, May 2013

The general food security trend in the county is fairly stable in the mixed farming (coffee, dairy, irrigation) livelihood zone but deteriorating in the mixed farming (food crop, cotton, livestock) and the marginal mixed farming livelihood zones. The most adversely affected areas are Kalawa, Nguu, Makindu, Kibwezi and Kathonzweni divisions (which all lie in the mixed farming livelihood zone).

Implementation of FFA activities in Kibwezi district by GoK, UNWFP and its co-operating partner in the district – the KRCS – began in October 2010 with 24,400 beneficiaries based on recommendations from the Kenya Food Security Steering Group's seasonal assessments of the short rains. These assessments were conducted in February 2010 and recommended the recovery operations in that part of the county. However, due to the deteriorating food security situation in the district because of frequent droughts, the number of beneficiaries was subsequently increased to 27,500, based on the recommendations of the short rains assessment which was carried out in February 2011, and finally to 54,600 beneficiaries after the long rains assessment carried out by the Kenya Food Security Steering Group (KFSSG) in July 2011.

Masongaleni lies within the marginal mixed farming zone and receives rainfall of between 300 mm and 400 mm per year. The location has a total area of 279.9 square kilometres, with small stock rearing being the dominant livelihood activity. River Athi acts as the northern edge of the location, bordering Kitui county, and is the main source of water during drought. The river is an average of 13 to 15 kilometres from most households, compared with the 5 kilometres to water sources during normal times. Some people, especially youths, have migrated to urban areas in search of formal employment as they find it increasingly difficult to survive the frequent droughts.

Within Masongaleni, a total of 2,127 households (or 61 per cent of households in the location) with 12,762 beneficiaries have been targeted. They are provided with a 50 per cent ration of food, estimated as the amount of food that can sustain the household's nutritional requirements for half a month. This ration is used because the FFA approach is designed so that the beneficiaries work for only 12 days in a month on project-funded work (three days a week) and are expected to utilize the rest of the month in producing food, or earning, to provide the remaining 50 per cent ration. This is intended to promote creativity and innovation in households and communities while at the same time avoiding the creation of dependency. According to the LAC framework, innovation is key for building adaptive capacity as communities may at times need to alter existing practices, resource use and behaviours, or adopt new ones, as social and environmental changes continue (Jones et al., 2010).

The Food for Assets approach as used in Masongaleni

In carrying out the FFA project activities, efforts have been made to ensure that community members identify their own projects. Projects selected are those that retain and create assets and invest in alternative livelihood options, including water harvesting technologies and the creation of on-farm assets on individual farms and communally owned land parcels, with technical support from government departments, the UNWFP and non-governmental organizations (NGOs).

Targeting. The community selected vulnerable households (12,762 beneficiaries) in collaboration with elected, gender-balanced relief committees (RCs). The selection process was done publicly (hence it was transparent and accountable), facilitated by the co-operating partner (KRCS), and based on vulnerability criteria determined by the communities themselves.

After the targeting process, an integrated analysis was carried out by various stakeholders such as local communities/beneficiaries, opinion leaders, women's and men's groups, stakeholder forums, government departments and other NGOs using participatory rural appraisal methodology. This focused on identifying the extent to which poverty and weather extremes affect the communities.

Project identification in Masongaleni through participatory rural appraisal. Planning meetings were organized between people in each cluster and a team from KRCS and district project steering committee (DPSC) representatives where the purpose of the proposed project was explained and the communities gave their views on problems of water shortage, crop production, health, sanitation, land degradation and markets, among other things.

Table 9.2 Breakdown of Food for Assets project planning units in Masongaleni location

Sub-location (all in Masongaleni)	*Total population (2009 census)*	*Final distribution point name/ planning unit*	*FFA beneficiaries targeted through community-based targeting and distribution*
1 Masimbani	6,586	Masimbani	1,980
		Makutano	2,112
2 Ulilinzi	6,742	Kithyululu	2,094
		Utini	1,278
3 Ndauni	5,982	Ndauni	3,570
4 Kyanguli	1,750	Kyanguli	1,728
Total	*21,060*		*12,762*

Note: The numbers in the final column were arrived at based on recommendations from KFSSG on the percentage of needy people in the district. The Kibwezi district steering group (DSG) then allocated the various percentages for the sub-locations/final distribution points (FDPs) based on the poverty levels in the areas.

The following gives an example of the project selection process for Cluster 1 (Kithyululu, Utini and Kyanguli FDPs):

1. A public meeting was held in Ulilinzi market, Masongaleni location, consisting of provincial administrators (chief/sub-chiefs), the KRCS, DPSC members and 66 community members from the three FDPs.
2. A planning team of 12 community members (four male and eight female) was formed by the community members and given the mandate to formulate the project interventions.
3. This was followed by problem identification and prioritization where members ranked the problems according to priorities and determined their causes and possible solutions, as shown in Table 9.3.
4. Participatory catchment and village mapping exercises were then carried out separately by both male and female planning team members to capture the priorities of the different genders.
5. The planning team members then developed a catchment development map based on the constraints analysis. Members agreed that the construction of micro-catchments for in situ rainwater harvesting as well as terraces should be done at household level and ex situ rainwater harvesting structures, such as the external catchment systems and earth dams or pans, should be done on communal or public lands.

Interventions to increase resilience

A number of interventions have been identified and implemented under FFA in Masongaleni location following the project selection process explained above. Their contributions to building resilience are explained below, together with illustrative photographs.

- 925 hectares of land (30.2 per cent of Masongaleni location) have been treated and conserved with physical soil and water conservation measures/structures (e.g. terraces, stone and soil bunds, etc.) and zai pits. These structures have reduced environmental degradation by reducing slopes, thereby promoting infiltration rather than run-off and the conservation of soil nutrients, which contribute to improved food production. Demonstrations of zai pit micro-catchments, which have been used to plant maize crops, have been undertaken in primary schools with the communities' involvement and replication carried out at household level. School pupils are being used as ambassadors to transfer the skills to the household level. The adoption of this technology by beneficiaries has been quite good. The beneficiaries have adequately maintained the structures, thereby increasing their lifetime. In addition, replication has been clearly evident, with non-FFA project beneficiaries paying for the services of trained FFA technology resource people to lay similar structures in their farms.
- 25,552 cubic metres of water pans have been rehabilitated through de-silting, fencing and construction of regulatory and auxiliary facilities (cattle

Table 9.3 Problem prioritization and ranking in Kithyululu, Utini and Kyanguli final distribution points

Priority no.	*Problem*	*Causes*	*Possible solutions provided by the planning team*
1	Food insecurity	Inadequate rainfall Soil erosion Deforestation Planting of unsuitable seed varieties Poor farming practices Inadequate/poor storage facilities Lack of rainwater harvesting facilities	Soil and water conservation through terracing Afforestation/establishment of tree nurseries Construction of micro-catchments, e.g. zai pits Planting of certified drought-tolerant seed varieties Technical support from line ministries
2	Lack of pasture/ browse	Inadequate rainfall Planting of unsuitable grass varieties Larger herds of cattle per household	De-stocking/rearing small herds of livestock Planting of suitable grass varieties Afforestation Construction of terraces
3	Inadequate water availability for domestic and livestock use	Inadequate rainfall amounts Fewer water sources Silting of existing dams/ pans Lack of experience in rainwater harvesting	Improved rainwater harvesting Planting trees to reduce evaporation losses from surface water De-silting of existing water pans (13 in total)
4	High poverty levels	Low household income High illiteracy levels Lack of adequate ready markets Overdependency on relief	Involvement in income-generation activities, e.g. poultry raising Educating the community Liaising with relevant ministry for market linkages/ improvement
5	Poor road networks	Damage to roads Lack of feeder roads	Rehabilitation of damaged roads to connect the following centres: • Machinery – Wandei • Kithyululu – Yikivuthi – Yikitaa • Kyanguli – Masaku Ndogo • Mwoka junction – Utini
6	Deforestation	Charcoal burning Ignorance	Afforestation

Source: Participatory rural appraisal file from the KRCS

troughs and latrines) and have harvested and stored run-off and greatly improved water availability for domestic use and livestock production, especially during the dry season. Water is now available for an extra two to three months after the cessation of the rains (the rehabilitated Makio dam is shown in one of the photographs). Additional benefits include a reduction in the time spent by women in search of water. The construction of auxiliary facilities is important in ensuring good quality water for domestic use, otherwise the water is often polluted and muddy and may be of little use to households.

- 79,151 tree seedlings have been produced and distributed to farmers for planting to help reduce environmental degradation and improve diversification of food sources (e.g. mangoes, which were planted in October 2011 and are expected to mature after five years) and browse availability (e.g. the leucaena tree, which is planted in grazing areas as a browse for goats).
- Six lined farm ponds (200 cubic metres each) have been created for harvesting off-stream flow to provide water for micro-irrigation purposes and agroforestry in collaboration with the World Agroforestry Centre (ICRAF). Rope and washer pumps have been installed in the ponds for pumping the water and are being utilized for agroforestry and micro-irrigation purposes by the six target households. Micro-irrigation is being carried out in quarter-acre (0.1 hectare) plots where vegetables are being grown mainly for domestic use and dietary diversification, with the surplus being sold locally in the market centres. Through this initiative, other stakeholders have come in to provide further assistance to the communities; the National Agriculture and Livestock Extension Programme (NALEP) is planning to build 20 farm ponds within the location as part of its project entitled 'ASAL best practices site'.
- 1,882 cubic metres of gully land have been reclaimed as a result of check dams (see photograph overleaf) and other gully rehabilitation structures. Implementation is labour intensive and the availability of stones is also a challenge in some of the sites, as well as the high cost of wire mesh, so further take-up has been low.

Impacts observed

The impacts of the interventions included the following:

- Increased water supply (e.g. water pans, earth dams, farm ponds, etc.), which constitutes an essential livelihood asset for the communities. This has also led to improved groundwater recharge in the areas around the pans.
- Increased food and nutrition security. Communities are able to grow different types of crops (mainly sorghum, maize, water melons, cowpeas, green grams, vegetables and fruits) on the created assets, which include both micro-catchments and external catchment systems (e.g. zai pits, soil bunds or ridges, etc.). However, a key challenge to the improved crop production has been the lack of access to markets. Bumper harvests of green grams were realized in Masongaleni but the lack of favourable markets led to the farmers selling their produce to middlemen at low prices. Because of this experience, emphasis is being placed on value chain analysis (VCA) to link

Beneficiaries fetch water from the rehabilitated Makio dam during a dry season.
Credit: Swalha Njogu, KRCS

Beneficiaries pricking tree seedlings in one of the nurseries established in Masongaleni.
Credit: Leah Sang, KRCS

A fully established tree nursery. *Credit:* Leah Sang, KRCS

A gully control structure (loose rock check dam) reinforced with wire mesh. *Credit:* Joel Mutiso

producers with markets. VCA describes the full range of activities required in the production and sale of goods and services from conception through the stages of production, processing and retail to final consumers. VCA is also useful in helping smallholder farmers to improve their engagement at the local level, as well as national trade.

- Reduced environmental degradation through the construction of terraces (both stone and soil bunds), establishment of tree seedlings and tree planting (i.e. under agroforestry and reforestation) and gully rehabilitation practices. The community members have been able to plant appropriate tree species on their farms and the observed survival rate after the October to December 2011 short rains season was estimated at 50 per cent. This is expected to greatly enhance environmental conservation as well as providing a product. For example, those who planted mangoes expect to harvest the fruits, while leucaena trees are used as animal feed.
- Active involvement of men, women and youths in identifying their own needs, prioritizing and planning, and in the implementation and evaluation phases, which has enhanced project ownership and will ensure sustainability in the long run. The beneficiaries undergo extensive training on all aspects of project management by the line ministry officials; this includes training in areas such as leadership skills, group dynamics, entrepreneurship skills, book/record keeping and financial management. The beneficiaries are, however, given the freedom of choosing whether or not to form a community-based organization or join an existing one, as most of the actions require collective action. Training is often provided for a few selected community focal people (resource people), who are in turn expected to train other community members using locally accepted procedures and methods.
- Improved capacity of the target beneficiaries due to targeted and intensive training, which has enabled community members to plan and implement the projects as well as owning and maintaining the project assets that are created.

Reasons for success

Success has been assured because of the following:

- The process for designing the project implementation plan has been clear and efficient, although sometimes the timing of the rains has interrupted construction work (especially for water pans).
- The process is participatory and collaborative, involving all stakeholders from community to national level. In turn, this has allowed partners and communities to set their sights beyond recovery, substantially increase the size and scale of assets created, support existing and alternative livelihoods, and shift the measure of success from outputs to outcomes and impact. For example, in the construction of terraces, reporting is done not in terms of kilometres constructed but on the acreage put under crop production. The

community has been trained on how to provide data on production from random sample sites, which helps in showing the impact. Comparisons are also made between production on farms provided with terraces and those without the structures.

- Through the adoption of the value chain approach, the planning of future interventions is being carried out by adopting a 'market-led approach' (rather than the 'product-led approach' adopted initially) and value chain groups (VCGs) are being formed so that the farmers can be well versed in market dynamics. For example, a green grams VCG has been formed in Masongaleni to ensure that farmers growing the crop are not exploited after bumper harvests.
- A key success has been the enhanced capacity building at all relevant levels – starting at national and district level (DSGs) and trickling down to the community level (RCs). The capacity-building programmes incorporated within the FFA interventions have imparted necessary knowledge and skills from the national, district and community level and access to market information. Knowledge and skills include: project management training as well as resource mobilization skills, which led to the forging of partnerships with ICRAF in Masongaleni, at the national level; at the district level, the training of the district FFA co-ordinator, co-operating partner (CP) technical officer and DSG members on rainwater harvesting and management for improved crop and pasture production, which led to the district teams being able to offer adequate technical advice to Masongaleni communities on viable technologies; and, at the community level, the training of relief and project committees and resource people on the layout of structures, which led to them taking the initiative, as well as leadership skills and group dynamics to ensure that they maintained harmonious working relationships between the members.
- Besides training and technical expertise, the national- and district-level capacity has been further enhanced through provision of capital assets such as vehicles and extra funds for facilitation from GoK and UNWFP, which has led to improved project implementation through the availability of continuous support to the beneficiaries, and therefore sustainability.

Challenges to Food for Assets implementation

In spite of the many benefits, FFA has a number of weaknesses that need to be addressed for it to support or qualify fully as a CBA strategy.

- Increasing rainfall variability, which results in frequent and prolonged droughts and is attributed to the changing climate, is one of the main challenges to realizing the benefits and impacts of FFA – rainfall is the 'killer bullet', since the performance of most of the FFA activities are rainfall dependent. Without adequate rainfall, there will be no adequate water or moisture available for plant growth (crops, pasture, trees and fruits) or for domestic and livestock use (water pans). From the choices of possible interventions in Table 9.3, ranging from terracing, zai pits, de-silting of pans and improved rainwater harvesting to planting of drought-tolerant varieties, it can be said that attention is being paid to conserving and maximizing benefits from the limited available water, which is key for adaptation in this semi-arid area. This may be strengthened further through the use of both short- and long-term climate projections.

- In some cases, inadequate supplies of certified seed of drought-tolerant crops have led farmers to plant unsuitable crop varieties. However, concerted efforts have been put in place through the Ministry of Agriculture and NGOs aimed at sensitizing communities to the importance of the adoption of drought-tolerant crops, specifically certified seeds, as well as to the importance of timely planting.
- Inadequate tools have affected the project implementation schedule as there are more people than available tools. Communities have had to share tools by working in shifts.
- Some FFA projects (for example, manually created water pans) are very small (less than 3,000 cubic metres), which limits their benefits and impacts. This is understandable due to the limitation of manual labour. However, planning for the next phase of FFA includes the use of machine power in the construction of dams and pans so as to increase their volume, reduce drudgery and build longer-lasting structures.

Lessons learned

Some of the lessons learned so far include the following:

- The forging of partnerships, for example with ICRAF, is a holistic and integrated approach which helps ensure sustainability. The inclusion of different partners in FFA projects is vital for access to additional inputs, whether of capital, knowledge or information.
- Strengthening integration of the value chain approach into FFA could significantly improve sustainability and help in the choice of crops and levels of production.
- Collective action is useful in creating assets for adaptation and ensuring that benefits are shared
- The use of climate information could improve success and sustainability of interventions in the long run. So far, use is made of indigenous knowledge and seasonal forecasts that are short term and useful in managing current variability. Use of long-term climate projections could help in strengthening the adaptive capacity further.
- Even though the different sectors of society (i.e. men, women and youth) are represented in planning and decision making, the prioritized activities do not seem to align closely with the needs of youth, which results in their higher levels of migration to towns during drought compared with other groups (see the sections above).

Sustainability

The sustainability of this approach is achieved through these factors:

- A participatory approach is adopted that embraces a consultative process from proposal preparation to implementation, by involving a wide spectrum of stakeholders (i.e. communities, line ministries, co-operating partners and administration officers). This has enhanced ownership of the assets by the communities, ensuring adoption, maintenance, replication and upscaling. The clear roles set for each partner also ensure that there are no conflicts of interest, with each partner in charge of specific duties; for example, the provincial administration is in charge of mobilization and law enforcement, communities are in charge of project formulation and implementation as well as maintenance, CPs are in charge of supervision, technical DSG ministries are in charge of technical support, and the donor (UNWFP) is in charge of funding and material provisions.
- Larger water dams or pans are constructed that are more long-lasting and able to stand in as a reliable water source for communities between the rainy seasons. In most community-driven activities, only smaller pans (less than 3,000 cubic metres) are achieved due to the labour-intensive activity involved. However, FFA projects aim to build pans or dams between 3,000 and 15,000 cubic metres in size by engaging a larger workforce and allowing more time for construction (up to six months), as well as having technical designs and supervision from the Ministry of Water. This ensures that the water collected is able to last the beneficiaries a considerably longer period of time after the cessation of rains, and the community will continue to find a reason to maintain and repair the structures since they find them useful to their livelihoods.
- There is a reliable exit strategy in place. Community engagement processes are strong and accompanied by continuous capacity building. Through this, communities are able to carry out their needs assessments, lay out the structures, operate them, and maintain or repair them periodically. The setting up of strong community structures (community project committees) has also enabled communities to run and manage their own projects. The involvement of a wide spectrum of partners, especially line ministries, ensures follow-up of the interventions.
- Sustainable environmental compliance mechanisms are the joint responsibility of community members, the National Environmental Management Authority, technical GoK ministries and other partners. This ensures that interventions selected are environmentally friendly and sustainable.
- Gender and social practices are considered. The FFA approach actively involves men, women and the youth in identifying their own needs and then prioritizing, planning, implementing and evaluating progress to help ensure that the diverse needs of these groups are catered for, and to strengthen project ownership and sustainability.

Conclusion

Kenya's ASALs are expected to become drier, with climate change characterized by rains that are poorly distributed in space and time. Adaptation to climate change in these areas will require having flexible systems that take into account these possible climate futures.

The major focus of the FFA strategy has been to build assets to strengthen resilience among targeted households and communities. A review of FFA experience in Masongaleni shows that even though they are not specifically designed as adaptation projects, they more or less contribute to strengthening adaptive capacity at the local level as outlined in the LAC framework. In the study area, we have participatory planning and decision making; the coming together of institutions and the local community; and the sharing and use of experiential knowledge and technical support from government officers and other institutions. These combine to create assets that help households and communities widen their livelihood options and improve the health of the environment itself, and that are key in building local adaptive capacity.

Bringing in different resources, knowledge and expertise is important in strengthening adaptive capacity and care should be taken to ensure that relationships are managed well. The fact that the FFA activities are aligned with the national climate change response strategy and the sessional paper on sustainable development of ASALs provides the programme with supportive policies, and potentially with resources from government, especially for scaling up some of the successful activities or technologies.

Note

1 ALRMP is a community-based drought management project of the Kenyan government using a credit facility from the World Bank (www.ndma.go.ke).

References

Adger, N., Khan, S. and Brooks, N. (2003) *Measuring and Enhancing Adaptive Capacity*, New York, NY: United Nations Development Programme (UNDP) <http://ffa.kenyafoodsecurity.org>.

C-SAFE (2004) *Food for Assets: Adapting Programming to an HIV/AIDS Context*, Johannesburg: C-SAFE Learning Centre <http://reliefweb.int/sites/reliefweb.int/files/resources/40A2E8EBA0C28CF7C1256F1100411E2A-csafe-zam-16sep.pdf> [accessed 7 October 2013].

Eriksen, S. and Kelly, P. (2007) 'Developing credible vulnerability indicators for climate adaptation policy assessment', *Mitigation and Adaptation Strategies for Global Change* 12(4): 495–524 <http://dx.doi.org/10.1007/s11027-006-3460-6>.

GoK (2010) *National Climate Change Response Strategy*, Nairobi: Ministry of Environment and Mineral Resources, Government of Kenya (GoK).

GoK (2012) *National Policy on Sustainable Development of Northern Kenya and Other Arid Lands*, Sessional Paper No. 8, Nairobi: Ministry of State for Development of Northern Kenya and Other Arid Lands, Government of Kenya.

GoK and WFP (2010) 'Food for Assets guidelines for project implementation teams', Nairobi: Government of Kenya and World Food Programme Kenya.

Herrero, M., Ringler, C., van de Steeg, J., Thornton, P., Zhu, T., Bryan, E., Omolo, A., Koo, J. and Notenbaert, A. (2010) *Climate Variability and Climate Change: Impacts on Kenyan Agriculture*, Nairobi and Washington, DC: International Livestock Research Institute and International Food Policy Research Institute.

Jones, L., Ludi, E. and Levine, S. (2010) 'Towards a characterisation of adaptive capacity: a framework for analysing adaptive capacity at the local level', ODI Background Note, London: Overseas Development Institute <www.odi.org.uk/publications/5177-adaptive-capacity-framework-local-level-climate> [accessed 7 October 2013].

Orindi, V.A. and Ochieng, A. (2005) 'Seed fairs as a drought recovery strategy in Kenya', *IDS Bulletin* 36(4): 87–101 <http://dx.doi.org/10.1111/j.1759-5436.2005.tb00236.x>.

About the authors

Victor A. Orindi works as a climate change adviser with the Ministry of State for Development of Northern Kenya and Other Arid Lands. Previously he worked for the International Development Research Centre in the Climate Change Adaptation in Africa Programme and the African Centre for Technology Studies.

Daniel K. Mbuvi works as County Drought Coordinator for the National Drought Management Authority, Kenya. He holds a master's degree in land and water management (range management option) from Cranfield University in the UK. He has worked in development projects assisting communities in the drought-prone arid and semi-arid lands of Kenya since 1985.

Joel W. Mutiso works as County Drought Response Officer for the National Drought Management Authority on food security in Makueni county, Kenya. He holds a bachelor of science degree in agricultural engineering from Jomo Kenyatta University of Agriculture and Technology. He is a member of the Southern and Eastern Africa Rainwater Network (SearNet).

CHAPTER 10

Indigenous knowledge and experience in adapting to drought in Vietnam

Le Thi Hoa Sen and Dang Thu Phuong

This study was done in Quang Tri, Vietnam, where drought and climate change impacts have been experienced. The study aims to identify indigenous knowledge and experience of Quang Tri farmers in adapting to droughts. Interviews, group discussions, field observation and questionnaire surveys were used to collect data. Research results showed that there were numerous forms of adaptation to drought that people in the commune were practising, based on their experiences and indigenous knowledge. The indigenous knowledge and experiences applied rely on three sources, including the family's older generations; neighbours/friends; and agricultural officers. The first two were the dominant sources of information. The household's adaptation capacity of indigenous knowledge into production depended on economic conditions of the household, household labour force and household off-farm activities.

Keywords: climate change, indigenous knowledge, drought, traditional varieties and breeds, climate perception, Vietnam

In recent years, the global climate has changed and the changes are due to both natural phenomena and human activities (Dow and Downing, 2007). These changes are shown by more frequent and higher intensity disasters such as floods, droughts, storms and tsunami in recent years. These changes have impacted severely on social, economic and environmental systems and have affected prospects for sustainable agricultural and rural development (Fischer et al., 2002). According to the United Nations Development Programme, Vietnam is one of five countries considered the most vulnerable to climate variability and extreme weather events. Within the country, the central coastal area is one of the most vulnerable areas to typhoons, storm surges, flash floods, drought and saline water intrusion (Chaudhry and Ruysschaert, 2007).

As in other developing countries in the region, agricultural production plays a crucial role in the livelihoods of the Vietnamese people. Agricultural outcomes are determined by complex interactions among people, policies and nature (Nelson, 2009). More specifically, agricultural production is one of the sectors most vulnerable to climate change and has been affected by climate change (Oyekale, 2009) in terms of long-term changes in temperature and precipitation, or the frequency and intensity of extreme weather events

http://dx.doi.org/10.3362/9781780447902.010

(Bradshaw et al., 2004). The questions posed in this paper are: what are the impacts of climate change on agricultural production and how have farmers adapted, and/or can they adapt, to climate change?

According to Smit and Wandel (2006), adaptation is 'adjustments to enhance the viability of social and economic activities and reduce their vulnerability to climate, including its current variability and extreme events as well as longer term climate change'. Adaptation is an important component of climate change impact and vulnerability assessment and is one of the policy options in response to climate change impacts (Rao et al., 2007). One of the common purposes of adaptation analysis in the climate change field is to estimate the impacts of climate change scenarios; based on the nature of these impacts, humans can develop more effective adaptation strategies (Smit and Wandel, 2006). Besides, adaptation to climate change is essential to complement climate change mitigation, and both have to be central to an integrated strategy to reduce the risks and impacts of climate change (Fischer et al., 2002).

The relationship between indigenous knowledge and natural disaster and adaptation strategies has aroused more interest in recent years (ISDR, 2008). Indigenous knowledge is seen as the basis for promoting decisions at the local level in rural communities (Boko et al., 2007). Incorporating indigenous knowledge into climate change and risk reduction policies can lead to the development of sustainable and effective strategies for mitigation and adaptation. It also increases the resilience of communities to disturbance (Robinson and Herbert, 2001).

In recent years, Quang Tri has been increasingly affected by climate change effects such as droughts, storms, saline intrusion and floods; of these, drought is the most critical (CRD, 2009). Drought has heavily and negatively influenced the daily lives of local people and ecosystems. This paper focuses on climate change tendencies and impacts, as well as local adaptation options and the adaptive capacity of local people in relation to drought. The study contributes analysis that could prove useful for agricultural communities in the coastal areas of Quang Tri province as well as contributing to the 'Provincial Target Program to Respond to Climate Change', particularly as it can provide useful information for policy makers and policy-level planning. The study was conducted within the framework of the project 'Local communities in Quang Tri improve their adaptive capacity and resilience to climate change', implemented by the Center for Rural Development in Central Vietnam and the British charity Challenge to Change, and funded by the Finnish Embassy in Vietnam. The specific objectives of the study are:

- to study climate change (extremes and variability) tendencies and impacts on agricultural production in the study area, with a focus on drought;
- to document and analyse farmers' indigenous knowledge and experience in adapting to drought in agricultural production.

Research methodology

The study site

Trieu Van commune, a representative sandy commune of Quang Tri province, was selected for this study. The commune has an extremely high poverty rate (about 33.3 per cent in 2009) and is increasingly affected by climate change.

Data collection

Secondary information on extreme climate events, climate variability and impacts on agricultural production was collected. Other secondary data from Dong Ha meteorological station, including on rainfall and temperature, were also gathered for analysis.

The primary information was collected by using different participatory rural appraisal tools, including group discussion, in-depth key informant interviews, household interviews, and observation.

Two group discussions were conducted with key informants, who included commune leaders, representatives of mass organizations and experienced farmers. Each group included 10 to 15 participants. The discussions focused on climate extremes and variability over time, frequency and intensity, the impacts of drought on agricultural production, as well as farmers' adaptation strategies.

A semi-structured questionnaire was designed for household interviews. Fifty-nine out of 627 households in the commune were randomly selected for interviewing. The questionnaire had two main modules: information on the impact of drought on agricultural production, including the effects on land, water resources, productivity, product quality, production cost and diseases; and information on adaptation strategies of farming systems to drought, such as access to information on policies and adaptation options, indigenous knowledge information sources and adaptation processes.

Data analysis

Both qualitative and quantitative data collected were synthesized and inputted into SPSS and Excel software for analysis. Descriptive and inference analyses were the major processing methods in this study.

Results and discussion

General climate conditions of Quang Tri province and the study area

Quang Tri belongs to the central coastal region, where it lies on the border of South and North Vietnam. The province has 81 per cent mountainous land, 11 per cent lowland and 7.5 per cent sandy soil. Quang Tri has experienced most of Vietnam's climate disasters, and their frequency and intensity are greater in this region than elsewhere.

According to Dong Ha meteorological station's records, the average temperature of the province was 24.9 degrees Celsius during 1976–2008. Months with the highest temperature were June and July, and the coldest months were December to January (see Figure 10.1). From 1976 to the present, the hottest day reached 42.1 degrees (24 April 1980) and the coldest one was 9.4 degrees (2 March 1982).

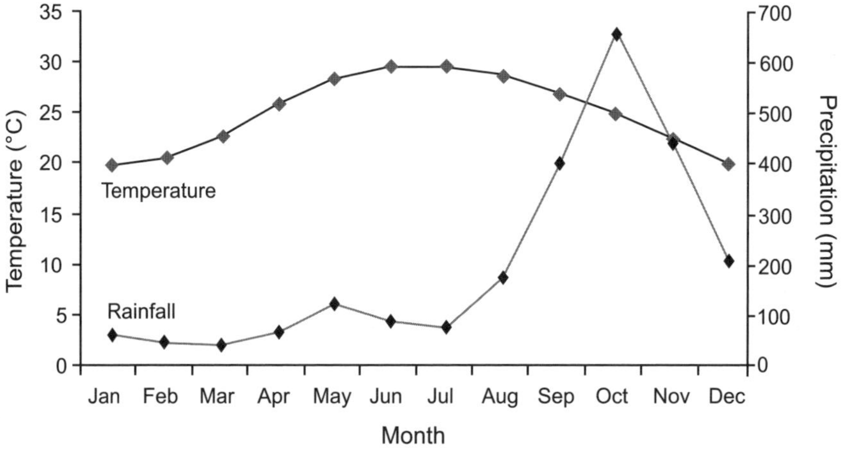

Figure 10.1 Average monthly temperature and rainfall in Quang Tri province, 1976–2008
Source: Dong Ha meteorological station

The average rainfall in Quang Tri province from 1976 to 2008 was about 2,300 mm per year. The highest total precipitation was 3,458.2 mm in 1980 and the lowest total precipitation was 1,424.5 mm in 1988. Over the course of a year, precipitation increased mainly from the end of August to November, reaching its highest level in October. The lowest precipitation was between January and April.

Basic socioeconomic figures of Trieu Van commune. Trieu Van commune has four villages: Village 7, Village 8, Village 9 and the Ecological village. In 2009, the commune had 2,398 people with 627 households.

At the time of the study, the main income sources of the population were crop production, livestock production, fisheries production, migration, small businesses and other off-farm activities. The crop production provided 65 per cent and livestock production 20 per cent of total family income. Besides the natural and social conditions of the research area being unfavourable, agricultural infrastructure was very poor. There was no irrigation system and no sea wall to protect against waves. This led to a shortage of water, together with saline intrusion in summer and waterlogging in winter.

Land resource and land use systems. The total natural area of the whole commune is 1,099 hectares, including 228 hectares of agricultural land (21 per cent). Of this land, 26 hectares (2 per cent) are affected by saline intrusion and 71 hectares (6 per cent) are waterlogged; 421 hectares are forest land (38 per cent), 139 hectares (13 per cent) are unused, and the rest of the land is used for other purposes. Small farm size is a key factor in agricultural production, affecting the efficiency of resource allocation and productivity.

The crop and livestock production system in the study area. The principal crops grown in Trieu Van commune were rice, sweet potatoes, cassava, peanuts, beans, water melon, onion and other vegetables. There were two main crops – a winter–spring season and summer–autumn. The cultivated area was larger for the winter–spring crop.

Rice and sweet potatoes were dominant crops in the study area. These two crops were mainly produced for home consumption. Sweet potato was a main feed source for livestock. Beans were an important crop in this area because of their nitrogen-fixing properties and therefore for improving soil quality, and they were also a valuable income source for farmers. On average, each household cultivated about 1,000 square metres of beans. Farmers practised intercropping beans with sweet potato and other annual crops. Due to the shortage of water, crops were less diversified in the summer season than in the winter, and the cultivated area of each crop was larger in winter.

Livestock production contributed significantly to the farm households' income in the study area. Pigs, cattle and poultry were major livestock components (Table 10.1).

Table 10.1 Livestock production of the surveyed households, 2009

Livestock	*Mean number (n=59)*	*Standard deviation*
Pigs	7.9	13.0
Sows	1.7	1.3
Cattle	0.5	1.1
Chickens	33.3	22.4
Ducks	10.1	39.4

Note: 59 households were surveyed.
Source: field survey, 2009

In 2009, almost all surveyed households kept pigs. Due to the weather conditions and feed resources, the pig breed used in this area is mostly the local breed called Mong Cai and F1 (Mong Cai x Yorkshire). Pigs were raised until three or four months old and then sold for cash. The average number of pigs in 2009 was 7.9 per household. Apart from pigs, cows were the main domestic animal. Cattle were considered a multipurpose animal. The most common cattle breed in the study area was the local breed known as yellow cattle. The average number of cattle per household was 0.5.

Poultry, including chickens and ducks, were kept around homesteads. Almost all farm households kept some poultry for home consumption and a few raised poultry for the market. Ducks were mainly raised in farm households where a reliable water resource was available. The average number of chickens and ducks per farm household was 33.3 and 10.1 respectively.

Water resources are crucial for aquaculture production. In Trieu Van commune, the fish pond was an important component in mixed farming systems. The role of the fish pond was not only to provide food for people or to sell for cash, but also to create suitable conditions for pig rearing and water for irrigation in the dry season. Only households with land with a high water table could have a fish pond.

Climate change and drought trends in the study area

Temperature changes. For quantitative analysis, the average temperatures recorded in Dong Ha meteorological station from 1976 to 2008 were analysed to see the change of temperature in Quang Tri province. Results of linear regression analysis of temperature data from 1976 to 2008 in Quang Tri show a rising trend for temperature; average temperature increased 0.0095 degrees per year, and the annual average temperature increased 0.3 degrees during the 33 years, although the R2 (coefficient of determination) was very low.

The qualitative analysis of people's perceptions was consistent with the regression analysis. It showed that about 89 per cent of interviewed farmers perceived long-term changes in temperature and 74.8 per cent of them thought that the temperature in Trieu Van commune had been increasing.

Precipitation change. Analysis of recorded data on precipitation in Dong Ha meteorological station from 1976 to 2008 showed that the annual rainfall and total rainfall in the rainy season and the dry season had not increased clearly during the 33 years (Figure 10.2). There is a large variability in the amount of precipitation from year to year.

An analysis of farmers' perceptions of changes in rainfall in the study area showed that 87 per cent of the respondents perceived changes in the rainfall pattern over the past 33 years, of whom 45.8 per cent noticed a decreasing trend in rainfall and that the rainy season was getting shorter. However, 18.6 per cent of interviewees thought that the rainfall had been increasing. Twenty per cent of the respondents noticed that the change was irregular. Rainfall distribution had also changed and did not follow the previous pattern, and therefore it was hard for farmers to forecast for their production planning.

Extreme climate events and drought. Extreme climate events in the study area include droughts, south-westerly wind, floods, cold spells and storms. Discussion with local farmers and commune staff indicated that storm

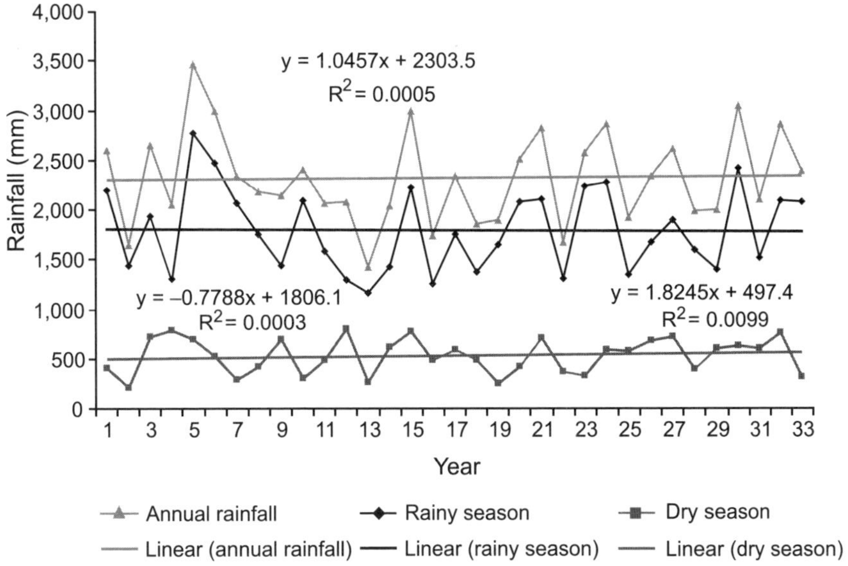

Figure 10.2 Annual rainfall and total rainfall in dry and rainy seasons in Quang Tri province, 1976–2008

Source: data from Dong Ha meteorological station

frequency and cold spells are decreasing. Droughts, floods and south-westerly winds are tending to increase but their intensity has been less over the past five years, while storms show an increasing trend in both frequency and intensity. Rainfall was estimated to be increasing during the rainy season and focused on certain times, meaning that the frequency and intensity of floods also increased.

Results of group discussions with local farmers indicated that drought duration was longer and varied from year to year. Commonly, the drought period was from April to the end of July, but within the last five years (2004–8), droughts had lasted from March to August and in some years lasted from February to August.

Increased air temperature and the intensity of south-westerly winds were considered the two main reasons for drought (96.6 per cent of surveyed households). The decreased rainfall in summer combined with high evaporation from the sandy soil were also reasons for drought.

Impacts of drought on agricultural production

Drought impacts on land resources in the study area. As discussed earlier, some climatic changes have occurred in the study area, notably drought. Impacts of increasing drought on agricultural production in the study area were discussed and the results showed that a reduction in cultivated area, water resources, crop and livestock productivity and quality were major impacts.

In group discussions with the village leader and commune staff, it emerged that cultivated areas in all three villages of the commune have been reduced due to water shortages and degraded soil. In Village 7, about 40 per cent of the total agricultural land could not be cultivated because of water shortages in the summer season. This figure was about 50 per cent and 60 per cent for Villages 8 and 9 respectively. The majority of crops were cultivated in the garden or areas close to households where they could use water from wells or ponds.

Results of group discussions showed that drought has impacted on land resources through reducing soil quality and salinity intrusion. About 93 per cent of respondents noticed that the cultivated area had been reduced in the summer season. Farmers in the group discussion also indicated that crops grew slowly or could not develop due to the dry and hot soil surface during the summer.

Drought impacts on crop production in the study area. Crop production in the study area depended largely on rainfall because there was no irrigation system, so crop productivity was affected significantly by drought. Table 10.2 presents farmers' opinions about the impacts of drought on crop production. It shows that increasing drought had led to the development of pests and diseases in all crops except for cassava and local onion. Sweet potato was the most affected crop: 93.2 per cent of interviewed households experienced worm infestation. For rice production, *Kho Van* disease showed an increasing trend and about 85 per cent of interviewed households experienced this disease.

Table 10.2 Impacts of drought on crop production in the study area

Crop production	*Increase in pests and diseases (% yes)*	*Increase in pesticides used (% yes)*	*Reduction in crop productivity (% yes)*
Rice (n=51)	84.6	9.1	90.9
Sweet potato (n=59)	93.2	5.6	94.4
Peanut (n=34)	82.9	17.9	78.6
Bean (n=40)	32.6	20.0	73.3
Bitter melon (n=37)	86.0	67.6	32.4
Casaba melon (n=31)	90.9	10.0	83.3

Note: n is the number of households growing the crop.
Source: field survey, 2009

According to the commune chairman, there were several reasons for the development of pests and diseases in crop production in the area, but increasing drought was the major one. He also indicated that the development of a number of pests and diseases in recent years has made disease control very difficult. Therefore, drought not only impacted on crop productivity but also on the quality of crop products and on production costs.

Households reported that sweet potato productivity had been affected by drought more than other crops. The development of wormy disease affecting the root systems of sweet potatoes led to a significant decrease in productivity of about 35.7 per cent compared with the spring crop and non-drought years.

For peanuts, respondents perceived that peanut productivity was reduced by about 30 per cent in 2009 due to increasing drought. In high temperatures and conditions of water scarcity, peanut often had *Chet Eo* disease, which cannot be controlled by any pesticides and leads to extremely low or no productivity. Casava crop and local onion crop were least affected by drought because these were drought-tolerant varieties. Bitter melon and casaba melon were affected much by increasing drought, as this leads to more pests and diseases for these two crops.

Drought impacts on livestock and aquaculture production in the study area. According to results of group discussions, drought directly and indirectly impacted on livestock. Direct impacts were increased disease, and reduced livestock health and scale of production. The indirect impacts were the reduction in feed resources and increase in production costs. Households perceived that impacts of drought have been increasing on livestock and aquaculture production. Figure 10.3 shows that cattle and fish were the most affected by drought. For fish, increased drought led to a lowering of the groundwater level and salt intrusion, and this had led famers to give up aquaculture production. For cattle, nearly 86 per cent of households experienced feed shortages due to increasing drought for nearly four months per year.

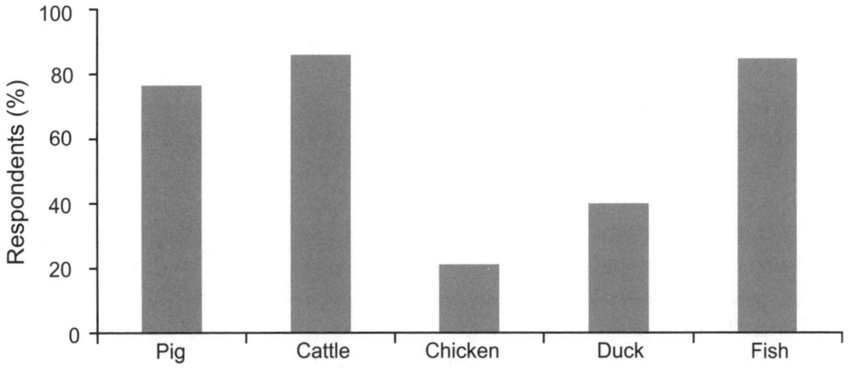

Figure 10.3 The impact of drought on feed resources (percentage of respondents answering yes)
Note: Numbers of respondents (n): pig n=59; cattle n=15; chicken n=53; duck=10; fish n=22.
Source: field survey, 2009

Livestock and aquaculture production were on a small scale in the study area. Their feed sources were mainly agricultural by-products and wild grass. High temperatures and shortage of water negatively affected the availability of grass. Pigs were the main livestock in the study site, and their main feed was sweet potato leaves and roots. As shown above, sweet potato was the crop that was most impacted by drought in the study area, so feed sources for pig production were gradually replaced by industrial feeds. In discussions about the industrial feeds, a majority of interviewees pointed out that it reduced labour costs for preparing feed for pigs and that pigs grew faster than with the traditional feeds, but the higher cost was the main barrier.

Regarding livestock health, all the surveyed households noticed that sudden changes of temperature led to increased livestock diseases, especially *Ga Ru* disease for chickens and *Bai Chan* disease for ducks. Sudden changes of air temperature also led to a change in the water temperature in fish ponds and lakes, and therefore fish growth was seriously affected.

Indigenous knowledge and experiences adapting to drought in agriculture

Results of household interviews showed that several adaptation strategies to drought were adopted by farmers in Trieu Van commune: using local drought-tolerant varieties and breeds, using indigenous knowledge and experiences to diagnose weather conditions for cultivation and adjusting the cultivation calendar, and using indigenous cultivating techniques and livestock management.

Local drought-tolerant crops and livestock. Using drought-resistant, pest-resistant and disease-resistant varieties was the main adaptation option for drought and climate variability. These varieties commonly have short growing periods and require less water for growth than others. They are normally harvested before the peak of the drought comes, thereby avoiding the effects of increasing drought. *Rit* was the only local rice variety and was characterized by high capacity to withstand salinity and drought, but it has a long growing period and low productivity. The area for cultivating *Rit* has been reduced significantly and introduced rice varieties (HT1 and *Khang Dan*), which have short growing periods and higher productivity, have replaced *Rit*. High-yield varieties have been preferred by farmers because they bring high productivity that meets the farmers' food and feed demands. Increasing crop yields have also been prioritized by the local government. The concern about using crop varieties that can adapt to climate change adaptation appeared after the concern about using high-yield crop varieties. The local government had policies and incentives to encourage the farmers to use high-yielding crop varieties, which had resulted in a loss or reduction in the use of varieties that are more adaptive to climate change adaptation although people were aware of the effects of climate change. Due to the impact of climate change, and to recent government awareness-raising on climate change impact and adaptation, both farmers and local government are now aware of the importance of crop varieties adaptive to climate change. The local government has already issued some advice encouraging farmers to plant such local varieties.

Unlike rice varieties and rice production, sweet potato production was not much influenced by advice from local government and sweet potato products are mainly for livestock production, so the production system has not changed much over time. Almost all surveyed households applied drought-tolerant varieties including *Tam Ky*, *Khoai chia* and *Khoai moi*. These varieties have been grown in the sandy areas for many generations and can only be found in this commune. All respondents revealed that over time these sweet potato varieties have increased their tolerance and have adapted well to the changing weather conditions in the area. They have a higher yield than other, newly introduced varieties.

For livestock production, results from group discussion showed that using local breeds was the major adaptation option for drought. Especially in pig production, the *Mong Cai* breed was the main local sow and was kept by almost all households. The F1 crossbreed between *Mong Cai* and *Dai Bach* (Yorkshire) was popular for raising for market. F1 crossbreeds had a high capacity to adapt to climate variation and drought and grew well with the local feeds, and its meat always had a better market price than that of the other breeds. For cattle, more than 72 per cent of interviewed households had a local breed called yellow cattle. However, local breeds of cattle, pig and poultry commonly have lower productivity than exotic breeds. The policy of the government on livestock production from now until 2020 encourages intensive production system with the use of exotic breeds. The government also gives incentives for farmers if they apply intensive livestock production systems. This may make the local breeds, which can adapt to climate change, difficult to use in the livestock production system. To increase the adaptation capacity of the livestock production system, the government should adopt policies and incentives encouraging the use of local breeds.

Indigenous knowledge of diagnosing weather conditions. Adjusting the seasonal calendar plays a crucial role in adaptation to climate change (Smit and Skinner, 2002). In the study area, commune leaders and experienced farmers had made adjustments. Based on their cultivation experience and indigenous knowledge of weather forecasts and the diagnosis of extreme climate events, the calendar was adjusted to avoid or minimize the effects of early floods, to take advantage of soil moisture before the rainy season was over for summer crops, and to avoid extreme cold and floods for winter crops. Leaders of the commune as well as interviewed households reflected that adjusting the cultivating calendar for agricultural production was an important adaptation option. Every year the local government (provincial and district departments of agriculture and rural development) make a seasonal calendar, which is the result of consulting many stakeholders and departments involved in agricultural production. The seasonal calendar has a legal status, which all individuals and organizations have to follow. Local staff were trained to make, adjust and implement the seasonal calendar. The local government also has policies to compensate partially for losses in agricultural production due to climate change and natural disasters, provided that farmers follow the seasonal calendar.

Cultivation techniques and livestock management. According to group discussion and household interviewing, various indigenous techniques have been applied in agricultural production to adapt to increasing drought in the study area, such as increasing the level of manure application in crop production; mulching; changing the crop structure or cropping patterns; trapping insects manually; and changing land preparation techniques. The results of data analysis showed that about 90 per cent of interviewed households increased the level of manure applied in crop production to improve soil fertility, to enhance the soil's capacity for retaining water and to moderate evaporation in the summer season. This was a low-cost adaptation option. Another technology that most farmers in the study area practised to keep soil moisture was mulching. Farmers made use of crop by-products, seaweed or grass to cover crop plots. Besides reducing water evaporation, mulch produced organic matter that retained nutrients for plants. Mrs Hoa in Village 9 stated that 'Before, most people carried almost all crop by-products home and dried them for an energy source for cooking. Nowadays, because of drought becoming severe in this area, the majority of people are now using crop residues to mulch their crops as a protection against drought.' In the study area, mulching was applied for three main crops, including sweet potato and bean, casaba melon and bitter melon. Farmers used seaweed to cover bitter melon roots, as it retains moisture and water and provides green manure.

People have been changing cropping patterns in the study area, to crops requiring less water or to crops tolerating drought: for example, changing from paddy rice to local sweet potato or casaba melon or peanuts. The initiative for change has come from some key farmers or outside groups. The effectiveness of changing the cropping pattern in terms of adapting to climate change and improved productivity convinced other farmers in the community to try doing the same. The local government also issues policies to encourage farmers to change their cropping pattern and gives some incentives for this.

In livestock production, there were many options practised by farmers to adapt to drought. Changing the design of animal houses was one; in order to moderate the impacts of high temperatures and south-westerly winds in the summer season, farmers designed a pigsty with higher walls and a high roof. About 55 per cent of surveyed households applied this option. Cattle sheds were designed more simply and at lower cost by using local materials. Storing feeds and utilizing supplementary feeds were the other adaptation options. Feedstuff was made by processing agricultural and fisheries by-products during the harvesting period or it was purchased on the markets whenever prices were low and then stored. More than 90 per cent of interviewed households applied one or both methods. The remaining households normally sold their pigs at any price when they faced feed constraints.

Factors affecting the use of indigenous knowledge to adapt to climate change in agricultural production

Previous analyses showed that there is a variety of indigenous knowledge and experience that farmers in Trieu Van commune have been practising to limit the impacts of climate change. It is likely that using indigenous knowledge has been hindered by factors such as the economic condition of the household, introduced technologies, the household labour force and the influence of authority figures. Government policies and the awareness and capacity of policy makers and technical staff are other factors. Indigenous crops and livestock commonly have lower productivity and a longer lifecycle than introduced ones. Introduced technologies were reported by discussion groups as the major barrier to adopting indigenous knowledge. Indigenous techniques such as mulching and applying manure to crops often need more labour than applying inorganic fertilizer. Thus, households with larger farm sizes and less labour are likely to be discouraged from practising indigenous techniques.

Among the most influential factors was the approach of the related technical departments and the helpfulness of the staff of local authorities. Indigenous knowledge is often ignored, as local authority staff prefer to introduce advanced techniques to farmers rather than relying on the 'traditional and backward' methods. There have been no specific policies to encourage people to keep using indigenous knowledge and practices, particularly in terms of conserving the indigenous crops and animals suitable for poor and small-scale farmers. In addition, successful indigenous adaptation models have not been considered suitable to upscale or integrate into the development plans of the local area due to the less convincing economic benefits, and due to limited awareness and therefore willingness of local staff. Support from local authorities and related departments, especially extension departments, and appropriate policies will be required to assist farmers in improving their adaptive capacity to drought in particular and to climate change in general.

In Quang Tri, there are non-governmental organizations (NGOs) working and supporting disadvantaged communities. Some of them are also supporting the study of indigenous knowledge and encourage its application in models at the community level. Some organizations aim to help local authorities to integrate adaptation models and indigenous knowledge into their economic development plans, with funding from government. However, this still requires much more time and longer-term projects and programmes to build capacity and awareness, and particularly to provide strong evidence to show the economic effectiveness of pro-poor interventions. For this reason, the continued support of NGOs in this field is critical, to help local authorities and technical departments, and even local poor farmers, to continue applying indigenous knowledge and practices to secure their crops and animals in the context of frequent disasters and changing climate.

There were three main sources of information on indigenous knowledge for farmers in the study area: family members (their own experience or that of older generations); neighbouring farmers and friends; and agricultural officers. Among these, family members and neighbouring farmers were the sources of information on indigenous knowledge that were relied on most. Agricultural extension officers, the staff of agricultural departments, co-operatives and even researchers prefer introducing new or 'advanced' technologies for farmers rather than suggesting traditional or indigenous knowledge.

In the study area, land use was decided mainly by the household head. Due to the infertile sandy soil, without irrigation or drainage systems, the local authority does not plan for any cropping systems, but they give farmers the right to make decisions by themselves depending on their own experiences and capacity to use their land resources optimally. This is the reason why farmers mainly apply indigenous knowledge. From their experience, new techniques are normally associated with high cost and high risk due to the poor soil and harsh climatic conditions.

The study found two patterns of adaptation process. The first was based on farmers' own experiences, leading to experiment, adjustment and then adaptation. The second involved learning best practices from others, testing or piloting these practices, adjustment then adaptation. The first was mainly followed by the better-off households, because they were more active and capable of developing experiments and testing adjustments than the others. If their experiments fail, they have other resources for their survival. The second pattern applied to all households, but especially for the poorer households, since they hesitate to take risks. However, in general these two adaptation cycles are interrelated.

Conclusion and recommendations

The results from statistical data analyses and farmers' perception in this research provided sufficient evidence to conclude that climate change was occurring and increasing drought was the most critical event. With increasing temperatures and the prolonging of the drought period, land and water resources were affected significantly. Cultivated land was reduced significantly in the summer season and the gap in the land use ratio between summer and winter seasons was increasing.

Crops, livestock and aquaculture production were increasingly affected by drought through the development of pests and diseases. Feed resources for livestock and fish became scarce because of decreasing productivity of crops and availability of grass.

There were numerous adaptation strategies being applied by farmers to limit drought impacts; most had been developed by indigenous knowledge and farmers' experiences and adopted autonomously.

The important messages from this research are as follows:

- In the context of climate change, it is important to document all effective indigenous knowledge that farmers have applied in the study area to disseminate to others.
- Authorities at different levels should formulate policies that can strengthen the capacity of local people to adapt to climate change in general and to drought in particular.
- Local government should promote anticipatory or planned adaptation for farmers by integrating climate change issues into rural development plans at all levels.
- There should be stronger support for poor farmers in the communities from civil society in co-operation with government agencies, which should help with documenting, maintaining, circulating and applying the indigenous

knowledge and practices. Appropriate policies to back this effort should be supported by those NGOs working in the province.

- More funding is needed for further research and application of drought-resistant models for the poor farmers at grassroots level in Quang Tri. This should come from international, national and local sources.

The change in agricultural livelihood practices may not be enough to adapt to climate change; there needs to be a profound transformation, not only in technical aspects but also in social aspects. Support from local government in terms of policies and incentives is important in driving the climate change adaptation process.

The use of indigenous knowledge in agricultural production is mainly for adapting to the changes of climate that have already been happening. There should be further research on the role of indigenous knowledge under different climate change scenarios, to assess the extent to which indigenous knowledge can address climate change impacts.

References

Boko, M., Niang, I., Nyong, A., Vogel, C., Githeko, A. et al. (2007) 'Africa', *Climate Change 2007: Impacts, Adaptation and Vulnerability. Contribution of Working Group II to the Fourth Assessment Report of the Intergovernmental Panel on Climate Change*, pp. 433–467, Cambridge, UK: Cambridge University Press <www.ipcc.ch/pdf/assessment-report/ar4/wg2/ar4-wg2-chapter9.pdf> [accessed 16 December 2013].

Bradshaw, B., Dolan, H. and Smit, B. (2004) 'Farm-level adaptation to climatic variability and change: crop diversification in the Canadian prairies', *Climatic Change* 67(1): 119–41.

Chaudhry, P. and Ruysschaert, G. (2007) 'Climate change and human development in Viet Nam', *Human Development Report* 46: 1–18.

CRD (2009) *Report on Livelihood Assessment of the Inland-Sandy Communities in Quangtri Province, Vietnam*, Hue City: Center for Rural Development in Central Vietnam (CRD).

Dow, K. and Downing, T. (2007) *The Atlas of Climate Change*, Berkeley, CA: University of California Press.

Fischer, G., Shah, M. and van Velthuizen, H. (2002) *Climate Change and Agricultural Vulnerability*, Johannesburg: International Institute for Applied Systems Analysis, World Summit on Sustainable Development.

ISDR (2008) *Climate Change and Disaster Risk Reduction*, Briefing Note 01, Geneva: United Nations Office for Disaster Risk Reduction (ISDR).

Nelson, G.C., Rosegrant, M.W., Koo, J., Robertson, R., Sulser, T., et al. (2009) 'Climate change: impact on agriculture and the costs of adaptation', Food Policy Report, Washington, DC: International Food Policy Research Institute <www.ifpri.org/sites/default/files/publications/pr21.pdf> [accessed 16 December 2013].

Oyekale, A.S. (2009) 'Climatic variability and its impacts on agricultural income and households' welfare in Southern and Northern Nigeria', *Electronic Journal of Environmental, Agricultural and Food Chemistry* 8(1): 13–34.

Rao, K.P.C., Verchot, L.V. and Laarman, J. (2007) 'Adaptation to climate change through sustainable management and development of agroforestry systems', *ICRISAT Open Access Journal* 4(1): 1–30 <http://oar.icrisat.org/id/eprint/2561>.

Robinson, J.B. and Herbert, D. (2001) 'Integrating climate change and sustainable development', *International Journal Global Environmental Issues* 1(2): 130–48.

Smit, B. and Skinner, M.W. (2002) 'Adaptation options in agriculture to climate change: a typology', *Mitigation and Adaptation Strategies for Global Change* 7: 85–114 <http://dx.doi.org/10.1023/A:1015862228270>.

Smit, B. and Wandel, J. (2006) 'Adaptation, adaptive capacity and vulnerability', *Global Environmental Change* 16(3): 282–92 <http://dx.doi.org/10.1016/j.gloenvcha.2006.03.008>.

About the authors

Dr Le Thi Hoa Sen is currently Vice Director of the Centre for Climate Change Study in Central Vietnam, an autonomous organization operating under the auspices of the Hue University of Agriculture and Forestry. Dr Sen works as a lecturer, researcher and consultant for NGOs and government agencies in climate change and agriculture. Dr Sen has conducted a number of research projects on indigenous knowledge and climate change adaptation in the agriculture of the northern, central and southern Mekong delta of Vietnam.

Dang Thu Phuong is the Climate Change Co-ordinator for the British charity Challenge to Change (www.challengetochange.org), with expertise on community-based disaster risk reduction and climate change adaptation initiatives in both urban and rural contexts of Vietnam. She is also a strong advocate for national climate change policies focused towards grassroots interventions benefiting the poor and most needy, including ethnic minorities.

Part Three
Conclusion

CHAPTER 11

Emerging lessons for community-based adaptation

Jonathan Ensor

The case study chapters in Community-based Adaptation to Climate Change: Emerging Lessons *provide a snapshot of current community-based adaptation (CBA) practice in different contexts and from different perspectives. While in each the focus is different, common themes emerge that suggest important lessons for practitioners, scholars and policy makers concerned with supporting communities in their continuing efforts to address the challenges of climate change and development. The value of participation and the nature of adaptive capacity in particular have recurred in the reflections of the practitioners who contributed to this volume, as summarized in this chapter. However, CBA also faces challenges in effective and equitable implementation, and in its scope and ambition. In the opening thematic chapters, the viewpoints of Yates, Reid and Cannon suggest three areas of neglect (politics of scale and technology, ecosystems and transformation) that CBA must come to grips with if it is to deliver on its potential and promises. These challenges are explored in the second half of this conclusion.*

Keywords: community-based adaptation, participation, adaptive capacity, politics, ecosystems, transformation

Lessons from practice

The contributors to this volume reflect the view of many researchers and practitioners in emphasizing the significance of participation in adaptation. Pradhan et al.'s (Chapter 6) focus on institutions in adaptation highlights the role that local or traditional ('informal') decision making can play in securing the representation or participation of those affected. Engagement with these community institutions in the Hindu Kush Himalaya region was found to render adaptation interventions more effective. For example, they document how in Mulkhow (Chitral, Pakistan) the community-managed water distribution system secures representation for villagers' interests and, through its relationship with an external agency, has underpinned the success of a modernization programme. This example also illustrates the potential for reform that resides within traditional, local decision-making structures. While community management systems are sometimes critiqued for their role in cementing inequitable power relationships at the local level, in Mulkhow the

http://dx.doi.org/10.3362/9781780447902.011

system is credited with presiding over the introduction of a more equitable distribution network as part of the modernization programme. As discussed elsewhere, these findings suggest that participation and equity are often best facilitated in context and from within local institutions, rather than imposed from without (Ensor, 2005; Ensor and Berger, 2009a; Hickey and Mohan, 2004).

In contrast, development, adaptation or modernization projects that bypass community institutions or lack transparency and accountability are found by Pradhan et al. to be less effective than the systems they replace (as in the case of a government-led intervention that overruled the community water management system elsewhere in Chitral, leaving villagers to repair failing irrigation channels). In some cases, the result has been to leave communities more vulnerable to climate-related hazards (as in the case of failing embankments for flood protection in Assam, India). Chambwera and Mohammed and Orindi et al. in Chapters 7 and 9 similarly associate the success of the projects they report on with the level of participation, in the former case linked specifically to the different social and ethnic groups involved in project implementation, and in the latter to the involvement of stakeholders from the community through to the national level. Indeed, all three chapters highlight the value of cross-scale interactions in adaptation actions, echoing the value placed on such collaborations to address environmental change by other researchers (Cash, 2006; Nelson et al., 2007; Osbahr et al., 2008). For Orindi et al. it was participatory approaches which engaged a wide range of stakeholders that 'enhanced ownership of the assets by the communities, ensuring adoption, maintenance, replication and upscaling', securing benefits for sustainability and broader impacts alongside local effectiveness.

The experiences reported by Koelle and Waagsaether in South Africa (Chapter 8) make a strong link between participation and adaptive capacity, showing how participatory action research (PAR) can 'enhance people's ability to learn together in the course of taking action to improve their situation'. PAR and similar approaches have a great deal to offer the development sector as it struggles to find ways of engaging with communities that meet their pressing, immediate needs while building the capacity to adapt to inevitable future uncertainty (Ensor, 2011). By contrast, Imbach and Prado's (Chapter 5) suggestion that the prevalence of externally originated projects and initiatives in their study is linked to a low capacity for local decision making and low project sustainability underlines the need to embed and develop adaptation projects locally if adaptive capacity is to be built. The ability to respond to the emerging challenges of climate change – which are liable to manifest in nonlinear and unpredictable, as well as incremental, changes to weather patterns – demands the capacity to learn and innovate; as recent research has noted, 'processes of inquiry, experimentation, and reflection are essential given incomplete knowledge about climate change' (Tschakert and Dietrich, 2010). Note, however, that Sen and Phuong's (Chapter 10) observations of autonomous adaptation in Vietnam reinforce the need for support to underwrite the risks of experimentation.

By marrying opportunities to gain knowledge and engage in experiments with working towards radical shifts in power relations and communities' sense of their own capabilities, PAR brings development focus onto key issues that define how and whether responses to climate change occur at the local level. Important cross-scale interactions, linking local people into networks of knowledge such as the metrological science and natural resource management communities, are fostered through the work described by Koelle and Waagsaether via a process that proceeds at a pace and with the sensitivity to ensure local ownership alongside external support. A system of 'community monitors' undertook research into specific questions of significance to the community, often collaborating with scientists to do so. This process helped shift power relations internally (with regard to gender, through the involvement of both women and men) and externally (linking across scales and with professions), encouraging 'farmers to explore their own questions in collaboration with scientists and other resource persons – and thus becoming the driver of their own investigation and learning processes'. In a similar way, the PAR process not only enabled farmers to access seasonal forecasts, but also led to farmers themselves recording, monitoring and ultimately demanding access to weather records, increasing their understanding of climate variability and their ability to communicate with scientists. Perhaps most significantly, Koelle and Waagsaether report the evolution of regular workshops, bringing the community together to share information, provide mutual support and stimulate shared learning. Together, these processes have enabled the community to plan adaptation strategies that respond to observed changes in extreme events and anticipated long-term changes. Notably, skilled, enabling facilitation and sustained support for the community's own development process over more than a decade have underpinned these important results for adaptive capacity.

Pradhan et al. offer an alternative perspective on adaptive capacity, focusing on the need to build 'synergistic' relationships between local and external institutions (that is, commonly understood formal or informal rules and norms). As they note, 'Adaptation cannot occur in a social vacuum – it needs to be supported by institutions and policies designed to enhance the adaptive capacity of local communities.' The concern here is therefore similar to that raised by Koelle and Waagsaether, but the lesson is different, relating to how those with responsibility to support communities can best achieve their aims through policy. Important links can be drawn here to the policy implications of Chambwera and Mohammed's chapter on the redistribution of responsibility (and costs) towards government and the private sector in order to scale up adaptation. For Pradhan et al., the success of support for adaptive capacity is found to depend on how effective policies and programmes are in relating to local institutions, and thereby integrating local needs and knowledge. In particular, the authors emphasize Agrawal's (2010) observation that public institutions – the instruments of policy – tend to collaborate with formal civic bodies at the local level, whereas effective local institutions often take the form of informal, traditional or community-based organizations.

Key components in Imbach and Prado's adaptive capacity framework are also bound up in institutions, with the ability to make changes linked to local perceptions of and reactions to climate change, and the processes that enable identification, preparation and implementation of actions. For Pradhan et al., the evidence suggests that support for adaptive capacity is best realized when policies work with local institutions in a manner that is targeted and responsive (rather than continuous and imposed). On the other hand, because responsiveness is a key issue for communities, programmes and policies that are supportive and enable communities to make changes can still be overtaken by local realities when market institutions 'respond more quickly to crop successes and failures that may be due to shifts in climate'. More broadly, as contributions from Mexico, Sudan, South Africa and Kenya in this volume illustrate, the role of the market (another institution) cannot be ignored in adaptation, and marketing (for example through co-operatives), market access and market information are crucial if livelihood diversification is to be considered. Orindi et al., for example, emphasize the use of a 'market-led approach' rather than a 'product-led approach' in assessing adaptation options. However, as Sen and Phuong note, indigenous crops may be better suited to local ecosystems and changing climate conditions, but have lower productivity or a longer lifecycle than introduced varieties, making them less attractive from an income and market-led perspective. In total, the focus on institutions and adaptive capacity in this volume helps community-based adaptation take a step forward in an under-researched field (Eakin and Lemos, 2010).

A conclusion explicitly reached by Chambwera and Mohammed and Orindi et al. in Chapters 7 and 9, as well as implicitly by others, is that sometimes what is needed for adaptation and adaptive capacity is good development. Chambwera and Mohammed's and Orindi et al.'s reviews of development projects show how both were able to provide outcomes that are significant in addressing the challenge of climate change. Chambwera and Mohammed find that the network of stakeholders linked to the local community through the development project are potential sources of adaptation resources and have an interest (direct or indirect) in building local adaptive capacity. They suggest that their work helps confirm Burton et al.'s (2006) view that a higher level of development is likely to produce greater adaptive capacity. Orindi et al. use the local adaptive capacity framework (Jones et al., 2010) to show how a Food for Assets (FFA) approach to development contributes to adaptive capacity, and find that key contributions are made in terms of access to technical advice, value chain analysis and broadening participation and collaboration. What remains in question, however, is whether it is possible (let alone desirable) to address adaptive capacity for climate change without an explicit focus on its multiple and context-dependent dimensions. As Orindi et al. acknowledge, a potential weakness of FFA is that it can build dependency, thereby killing off innovation and adaptive capacity. Elsewhere, the pitfalls of failing to explicitly address adaptive capacity even within adaptation projects have been documented (Ensor and Berger, 2009b), while Reid (Chapter 3) cautions against the labelling of development projects as CBA as, if they fail to deliver adaptation benefits, CBA itself may be discredited.

Challenges to community-based adaptation

The opening thematic chapters presented by Yates, Reid and Cannon offer up challenges to CBA. While supportive of CBA in principle, each suggests that there are important areas of neglect, where CBA has failed to engage consistently, undermining or overlooking the potential for delivering effective adaptation actions. These can be broadly summarized as falling into three interconnected areas, discussed in the following sections:

1. neglect of politics of technology and scale;
2. neglect of ecosystems;
3. neglect of transformation.

Neglect of politics of technology and scale

Yates's chapter challenges practitioners to take theory seriously and better understand the complexity of social relations and political power. Power and politics are not new development themes, but both are too frequently neglected in development practice in general and in adaptation in particular. In his contribution, Yates is concerned with two specific functions of power and politics – first, in relation to technology, knowledge and expertise. How is the knowledge generated on which adaptation actions are based? Whose interests are embodied in and served by technologies? As Yates points out, a technology is not neutral: its presence reflects political choices made in its development and promotion, and its utilization in a particular context will have political consequences – not least in terms of the distribution of resources, and the power to control the technology and its benefits. This is seen, for example, in the preference for and consequences of promoting 'advanced' technologies to farmers in place of indigenous methods (Sen and Phuong on Vietnam, Chapter 10). But this also leads to a second, related issue: recognizing that the uneven power and the complexity inherent in human systems mean that interventions often result in unintended outcomes.

Interventions – including adaptation policies and technologies – have an impact on actors who are not simply passive recipients. They change their actions in response to changing circumstances, often with unexpected results. As Yates's example of electric fencing to protect villages in Nepal demonstrates, both human and non-human actors in a system can change their behaviour (in this case, rhinos and community members in particular), with the result that decision making over the use and management of natural resources shifted dramatically, from the National Park and local government to community forest user groups. However, the introduction of the electric fence by the non-governmental organization (NGO) not only shifted decision making to community groups (that is, altering the institutions of forest and park management), but also blurred the neat distinction between the National Park, forest, local government and village scales, as villagers were left to cope with the tensions between the resources in the forest and the park. For them, the forest and National Park became one, and the local government faded into the background in the face of these challenges.

Yates echoes Pradhan et al.'s focus on how institutions relate to one another (how they 'interface'), but goes further by drawing attention to institutions as constantly changing as a result of social and political relations. Thus institutions are more slippery than simple definitions such as 'formal', 'informal', 'local' or 'traditional' may allow, in reality operating at or between different levels and in turn changing how we perceive those levels (or 'scales') in the process.

By examining the techno-politics of scale in Yates's Nepal example, it is possible to delve deeper into Pradhan et al.'s concern about the interface between institutions, and reveal how the outcome is an adaptation defined not by the best intentions of the NGO, but by how the networks of technology and practice carried out by the NGO intersected with the networks of knowledge, practices and norms surrounding the community.

Significant lessons emerge from these insights. As CBA starts to engage with institutions and engage in 'cross-scale' or 'multi-scale' approaches (including those examples of collaboration discussed above), Yates's message is that we need to be cautious in treating both scale and institutions as fixed or, indeed, easy to define and identify. Moreover, we need to recognize that adaptation actions have the potential for unintended consequences that extend to altering perceptions of scale and changing the opportunities for participation, representation and the decision-making rules (that is, the institutions) on which adaptation may rely. Participation is a feature of many of the case studies and, as discussed above, is recognized as an important factor in the success of adaptation. But the potential for powerful actors to define adaptation needs (and thus success) and for adaptation itself to shift the locus of decision making (and therefore the nature of participation) means that we must tread with care and, in particular, seek out the implications of adaptation actions from the perspective of those who are marginalized.

A real risk lies in the 'community' focus of CBA: like Cannon in Chapter 4, Yates warns against allowing 'community' to become a homogenizing idea or meta-narrative that conceals the complex realities of politics and power relations. Indeed, we need 'more nuanced understandings of communities as networks that are structured by unequal power relations and unequal access to knowledge, resources and decision making' (Yates, Chapter 2). Pradhan et al., for example, draw attention to how relationships between village leaders and local officials determine the support that communities can access, with political power manifested in those local leaders who are able to work across scales and define local adaptation priorities. But we also need to move our focus beyond the community and avoid the trap of scale and institutions: in Nepal, for example, did the NGO ignore the implications at the National Park scale in favour of a focus on the community?

These may appear to be difficult issues for practitioners to address, and they certainly challenge us to reassess the methods and assumptions upon which actions are based. In the end, complexity and political realities cannot be allowed to be sidelined, subconsciously boxed off as too difficult to cope with. It may mean that we need to re-skill and reassess the modalities of development, but to ignore the politics of technology and scale would be to ignore the responsibility that goes with adaptation practice. More positively, as Yates suggests, understanding the problem means that we have tools that can be employed. Techniques such as PAR (Chapter 8) – in which scientific research has to be shared with, demonstrate benefit to and ultimately be undertaken with communities – illustrate the steps that can be taken towards the resolution of unequal technology and knowledge resources. We can seek to understand 'the causes and scales of environmental injustice' by working alongside communities to identify whose knowledge and voices are marginalized in the complexity of development processes. Deliberate efforts to identify the context-specific nature of scale and technology politics suggest a need for forms of participation that re-cast interventions as co-development in which NGOs and communities are 'co-learners' (Koelle and Waagsaether, Chapter 8), enabling communities and those who seek to support them to make, in Yates's words, 'more informed choices about who benefits' from different kinds of adaptation actions. Power and knowledge differentials within and between scales, and the opportunities to manipulate scale and institutions that power and knowledge provide, can be discerned when we look for them and may be evident to the communities that are the focus of CBA. But understanding and responding to these realities will take a conscious effort on behalf of adaptation practitioners.

Neglect of ecosystems

Reid's contribution (Chapter 3) suggests that CBA has been guilty of overlooking a central component in people's lives: the natural systems on which we all depend. Ecosystems and biodiversity are on the front line of climate change, with ecosystem boundaries and species abundance already shifting as a consequence of changes in aggregate and extreme weather patterns, and profound, irreversible changes predicted as global warming reaches and stretches beyond 2 degrees Celsius. Ecosystem-based approaches (EbA) emphasize the need to protect and conserve biodiversity, but, as Reid notes, the Convention on Biological Diversity (CBD) definition has human adaptation at its centre. While CBA usually engages with natural resources as a central component of livelihoods and the mediator of climate impacts (as illustrated by all the case study chapters in this volume), EbA goes further by emphasizing the role of ecosystems in providing the raw materials necessary for adaptation (frequently overlooked by CBA, but see, for example, Sen and

Phuong's discussion of indigenous knowledge and resources in Chapter 10) and the role played by ecosystem services in securing human well-being (MEA, 2005), routinely overlooked by CBA.

Yates's chapter demonstrates that adaptation engages with complex problems because the act of intervening stimulates communities and other actors to reorganize and change their actions; the links between cause and effect can be difficult to discern, and the intentions of adaptation interventions do not translate into outcomes in a straightforward way. Reid makes a similar and important point in relation to ecosystems. CBA practitioners 'need to move beyond the perception of ecosystem goods and services as a set of static, finite natural resources, towards a fuller understanding of ecological complexity and interdependence' (Reid, Chapter 3). An ecosystem perspective highlights the links between different natural systems and the different roles they play (or 'services' they provide, such as pollination, or the regulation of climate, water quality or pests). The interconnectedness within ecosystems means that interventions need to be conscious of the full range of possible consequences and that, as in social systems, the complex relationships between components make it hard to predict the outcomes of actions that disrupt apparently isolated functions (such as digging wells or building dams for water access). Monitoring and learning to support adaptive management – that is flexible and responsive enough to respond to the unexpected consequences of human *or* climate changes – becomes critical (Cundill and Fabricius, 2009).

As with Yates's contribution on social systems, scale is an important factor in Reid's treatment of EbA. An ecosystem perspective draws attention to two scale dimensions – spatial and temporal. First, it may not be sufficient to think about only the local scale when communities depend on and are interconnected with ecosystems that cover much larger areas (such as a watershed or landscapes of connected habitats). Reid's point is that 'the potential benefits of CBA may be undermined if broader ecosystem processes and services are not considered in planning and implementation'. The ecosystem scale that, for example, connects multiple communities via rivers and run-off in a watershed region needs to be part of CBA thinking if the adaptations in one location are not going to be experienced as maladaptation in another. Equally, actions may be necessary outside the community scale (and beyond local administrative boundaries) to provide ecosystem services that benefit local people (as in the case of reforestation to improve flood water regulation). The second dimension – the temporal scale – relates to the different response times between human and natural systems. A community focus can implicitly limit the time horizon of CBA to that necessary for social, political or economic change. Yet climate change, and ecosystem responses to climate change, usually operate over much longer periods. Ecosystems that may appear static from a practitioner's perspective are in fact constantly changing and may respond to climate change long after the completion of a project or programme cycle, shifting, for example, the boundaries of crop pollinators and predators assumed to be stationary at the time when new crop varieties are introduced.

The lesson for CBA practitioners is that we need to see people as being located within linked social–ecological systems. This is a key challenge that places additional burdens on the institutions of CBA in particular. While, as Yates suggests, scales and institutions need to be understood to shift through the manipulations of knowledge and power, they also need to be seen more broadly, encompassing ecosystem space, connectivity and timescale. From this perspective, the risks that Yates discusses in multi-scale and collaborative approaches become an inevitable part of CBA as it seeks to integrate EbA and address the multiple interactions within and between human and ecological systems that are the context for adaptation. Moreover, if adaptive management is to succeed as a response to the complexity of this context, then it must become embedded in institutions that navigate multi-scale political and ecological change.

Neglect of transformation

In Chapter 4, Cannon follows Yates and Reid in challenging the CBA community to take context seriously. However, the concern here is broader, and the charge against CBA more pointed. For Cannon, 'CBA initiatives are currently too focused on adaptations to farming, and are not providing the basis for significant livelihood diversification at the local level'. Worse, CBA ignores the structural conditions that hold people in poverty and neglect background 'development' problems, calling into question 'the validity (and fairness) of adaptation measures for rural economies that are already often in crisis'. While there is evidence that CBA does engage with structural issues – for example, in Chapter 7 climate change in Sudan is recognized as overlying existing drivers of drought and conflict – Cannon's charge is to be taken seriously. What are we doing about those whose opportunities to adapt are so severely constrained or dependent on others that there is little realistic prospect of autonomous adaptation? What is to be done where climate change of 2, 3, 4 or more degrees renders existing livelihood practices unviable?

Cannon's suggestion (and, as he notes, it is not more than a hypothesis) is that the rural non-farm economy (RNFE) may be an overlooked adaptation option that can simultaneously reduce climate dependency, poverty, inequality and inequitable social relations, essentially sidestepping 'the rural power issues that at the moment maintain inequality and poverty'. The RNFE is defined as 'livelihood activities ... based in rural areas or pursued by people who are from households that are mainly rural-based, which do not involve direct agricultural production in crops or livestock' (NRI RNFE Project Team, 2000). Strong links therefore exist between livelihood diversification – a central pillar of CBA and recognized as an adaptation strategy in, for example, Pradhan et al.'s study of the Hindu Kush Himalaya region in Chapter 6 – and Cannon's proposal. However, while CBA operates with a local focus and frequently adopts a model of crop rather than livelihood diversification, work on the RNFE demonstrates that when conditions are right, a widespread transformation of livelihoods can be achieved away from those directly dependent on natural resources.

Yet, while there is much that we know about the RNFE, understanding (not to mention achieving) the right conditions is not a straightforward task, as Cannon's chapter makes clear. One key lesson is that large-scale investment by government is a central component, suggesting the potential for adaptation funding to act as a catalyst for livelihood transformation. Crucially, there is a role for local government in fostering an appropriate environment for investment, for example by bringing together entrepreneurs, community-based organizations and credit institutions. There is also a role for national government in identifying and making strategic investments that support economic activities in rural areas. There is support for a focus on diversification and the market in the case study chapters. As noted, Pradhan et al. highlight the responsiveness of the market to climate change and reflect that, rather than climate change awareness, it is markets and government policy that have an impact on adaptive capacity. Chambwera and Mohammed, on the other hand, directly link the potential of diversification to investment, and reflect Cannon's findings in stressing that 'involving the private sector and availing communities with loans to engage in livelihood diversification and income-generating activities' can play a role in scaling up. But the role of the state remains paramount: as Cannon stresses, the success of the RNFE in China is associated with a '*very significant role for state intervention*' rather than untrammelled market forces (his emphasis).

Other authors have critiqued CBA for failing to engage with broader structural issues, power relations and the challenge of transformation. For example, Dodman and Mitlin (2013) suggest that CBA risks losing credibility unless we can 'recognise that there is also a need to deal with institutionalised power relations above the level of the settlement, and this requires community structures that enable local groups to work together to represent their interests within these political structures'. Similarly, Ensor (2011) proposes that building on the lessons and practices of rights-based approaches to development holds the potential to transform social relations and power structures in favour of the poor through politicized action and reworked networks of accountability. Cannon looks for transformation from a different perspective, seeking modalities to support a rapid economic shift – in livelihoods – and thereby away from inequitable systems of tenure and power that block access to land and water and lead into landlessness and indebtedness. He proposes that diversification into the RNFE has been overlooked in the focus on CBA, and in particular that, by moving attention to the community scale, the potential to support a widespread transformation of livelihoods via the RNFE has not been realized. Most persuasively, there are clearly circumstances in which rural resource-based livelihoods will cease to be tenable due to a combination of climate change and underlying structural vulnerabilities, and as yet CBA has had little to say about how households and communities so affected can be supported. Yet concerns remain about whether a top-down approach to adaptation can engage with current livelihood diversification strategies and preferences, and with how power (and the techno-politics of scale) can be overcome to yield equitable outcomes in the face of what would be a rapid and profound redistribution of resources and wealth.

Participation – a central pillar of CBA – is frequently held out as the solution to such issues. But, as Dodman and Mitlin (2013) observe, participatory approaches have their limitations; in particular, they have, in the main, failed to engage with political change. Insofar as transformation is seen as a challenge of tackling social relations and the political structuring of power, social movements of 'politically aware and active organised citizens' are what most readily come to mind – not mainstream, locally focused participatory approaches to development (ibid.). As concluded elsewhere (Ensor, 2011), movements and alliances of the poor, such as Shack/Slum Dwellers International or the international peasant movement La Via Campesina, and the organizations and activists that make up their membership, are among those leading calls for the poorest to play a defining role in their own futures. In many cases, they have articulated specific demands for action on climate change adaptation (for example, Alliance for Food Sovereignty in Africa, 2009).

This is a view of adaptation that builds links between adaptive capacity and transformation, through political action that transforms social relations in ways that open up opportunities to meet future climate uncertainty. Development actors who recognize the long-term and profound challenges of climate change need to ask themselves how they can best support these calls for change and secure, in the language of rights-based approaches, the ability to make sustainable claims against those with the responsibility to support adaptation. Yet Cannon's chapter (Chapter 4) also invites us to think more broadly about transformation and to take seriously the need for investment, incentives, governance and a market that can catalyse and support alternative livelihoods or (in the case of urban communities in particular) infrastructure. And, as the engagement with ecosystems discussed above suggests, seeing communities as embedded in linked social–ecological systems means that transformation will need to account for ecological as well as socio-political sustainability. Integration of these themes – of equity, economy and ecology – is at the heart of the challenge of transformation and CBA.

References

Alliance for Food Sovereignty in Africa (2009) 'Alliance for Food Sovereignty in Africa (AFSA) challenges African leaders on climate change', Bole Declaration, 25 November 2009, Addis Ababa, Ethiopia.

Agrawal, A. (2010) 'Local institutions and adaptation to climate change', in Mearns, R. and Norton, A. (eds), *Social Dimensions of Climate Change: Equity and Vulnerability in a Warming World*, pp. 173–98, Washington, DC: World Bank.

Burton, I., Diringer, E. and Smith, J. (2006) *Adaptation to Climate Change: International Policy Options*, Washington, DC: Pew Center on Global Climate Change.

Cash, D.W. (2006) 'Countering the loading-dock approach to linking science and decision making: comparative analysis of El Niño/Southern Oscillation (ENSO) forecasting systems', *Science, Technology & Human Values* 31(4): 465–94.

Chambwera, M. and Stage, J. (2010) *Climate Change Adaptation in Developing Countries: Issues and Perspectives for Economic Analysis*, Environmental Economics Discussion Paper, London: International Institute for Environment and Development.

Cundill, G. and Fabricius, C. (2009) 'Monitoring in adaptive co-management: toward a learning based approach', *Journal of Environmental Management* 90(11): 3205–11 <http://dx.doi.org/10.1016/j.jenvman.2009.05.012>.

Dodman, D. and Mitlin, D. (2013) 'Challenges for community-based adaptation: discovering the potential for transformation', *Journal of International Development* 25: 640–59 <http://dx.doi.org/10.1002/jid.1772>.

Eakin, H. and Lemos, M.C. (2010) 'Institutions and change: the challenge of building adaptive capacity in Latin America', *Global Environmental Change* 20(1): 1–3.

Ensor, J. (2005) 'Linking rights and culture-implications for rights-based approaches', in Gready, P. and Ensor, J. (eds), *Reinventing Development?: Translating Rights-based Approaches from Theory into Practice*, London: Zed Books.

Ensor, J. (2011) *Uncertain Futures: Adapting Development to a Changing Climate*, Rugby: Practical Action Publishing <http://dx.doi.org/10.3362/9781780440392>.

Ensor, J. and Berger, R. (2009a) 'Community-based adaptation and culture in theory and practice', in Adger, W.N., Lorenzoni, I. and O'Brien, K. (eds), *Adapting to Climate Change: Thresholds, Values, Governance*, Cambridge: Cambridge University Press.

Ensor, J. and Berger, R. (2009b) *Understanding Climate Change Adaptation: Lessons from Community-based Approaches*, Rugby: Practical Action Publishing <http://dx.doi.org/10.3362/9781780440415>.

Hickey, S. and Mohan, G. (2004) 'Towards participation as transformation: critical themes and challenges', in Hickey and Mohan (eds), *Participation: From Tyranny to Transformation?*, pp. 3–24, London: Zed Books.

Jones, L., Ludi, E. and Levine, S. (2010) *Towards a Characterisation of Adaptive Capacity: A Framework for Analysing Adaptive Capacity at the Local Level*, London: Overseas Development Institute.

Millennium Ecosystem Assessment (2005) *Ecosystems and Human Well-being: Biodiversity Synthesis*, Washington, DC: World Resources Institute.

Nelson, D.R., Adger, W.N. and Brown, K. (2007) 'Adaptation to environmental change: contributions of a resilience framework', *Annual Review of Environment and Resources* 32(1): 395–419 <http://dx.doi.org/10.1146/annurev.energy.32.051807.090348>.

NRI RNFE Project Team (2000) *Policy and Research on the Rural Non-Farm Economy: A Review of Conceptual, Methodological and Practical Issues*, London: Natural Resources Institute (NRI), University of Greenwich.

Osbahr, H., Twyman, C., Adger, W.N. and Thomas, D.S.G. (2008) 'Effective livelihood adaptation to climate change disturbance: scale dimensions of practice in Mozambique', *Geoforum* 39(6): 1951–64 <http://dx.doi.org/10.1016/j.geoforum.2008.07.010>.

Tschakert, P. and Dietrich, K.A. (2010) 'Anticipatory learning for climate change adaptation and resilience', *Ecology and Society* 15(2): 11 <www.ecologyandsociety.org/vol15/iss2/art11/>.

About the author

Jonathan Ensor is a lecturer at the Centre for Applied Human Rights, University of York, where he undertakes research, teaching and practice focused on the environment, development and human rights. He has written widely on community-based adaptation and the relationship between climate change and development practice, including *Uncertain Futures: Adapting Development to a Changing Climate,* published by Practical Action Publishing in 2011.

Index

Note that f, t and box after page numbers indicates material found in figures, tables and boxes respectively.